Next-Generation Computational Drug Discovery

This book presents the fundamental principles, contemporary methodologies, and practical applications that define modern computer-aided drug design. It explores the integration of Cryo-EM structures, AlphaFold predictions, and homology modeling to expand structural insight into biologically relevant targets. In addition, it covers molecular docking, molecular dynamics simulations, enhanced sampling approaches, and free-energy calculations for robust hit-to-lead refinement. This book further examines in silico Absorption, Distribution, Metabolism, Excretion, and Toxicity (ADMET) prediction, quantitative structure–activity relationships, machine learning-based modeling, and network pharmacology frameworks applied across therapeutic areas, including oncology, infectious diseases, neurodegenerative disorders, and metabolic and cardiovascular conditions.

Key Features:

- Explores foundational and advanced Computer-Aided Drug Design (CADD) techniques, including molecular docking, molecular dynamics simulations, in silico ADMET prediction, and network pharmacology
- Integrates Cryo-EM, AlphaFold, and homology modeling to enhance structure-based drug discovery
- Details machine learning applications for pharmacokinetic profiling, toxicity prediction, and lead prioritization
- Discusses computational workflows for multi-target drug design across major therapeutic domains
- Presents systems biology perspectives and translational case examples linking computational discovery to experimental validation

This book is intended for researchers and students in pharmaceutical sciences, computational biology, bioinformatics, and related disciplines.

Next-Generation Computational Drug Discovery

Edited by
Mohammad Kalim Ahmad Khan
and Mohammed Tarique

CRC Press
Taylor & Francis Group
Boca Raton London New York

CRC Press is an imprint of the
Taylor & Francis Group, an **informa** business

Designed cover image: Shutterstock

First edition published 2027
by CRC Press
2385 NW Executive Center Drive, Suite 320, Boca Raton FL 33431

and by CRC Press
4 Park Square, Milton Park, Abingdon, Oxon, OX14 4RN

CRC Press is an imprint of Taylor & Francis Group, LLC

ISBN: 9781041140429 (hbk)
ISBN: 9781041140436 (pbk)
ISBN: 9781003672760 (ebk)

DOI: 10.1201/9781003672760

Typeset in Sabon
by codeMantra

Contents

Preface

The contemporary landscape of drug discovery has undergone a decisive transformation driven by the convergence of structural biology, computational chemistry, systems pharmacology, and artificial intelligence. Classical empiricism, which was the dominant paradigm of therapeutic development, has progressively yielded to rational, data-driven design frameworks that interrogate biological systems at atomic, molecular, and network scales. The contributions assembled in this volume collectively reflect this transition, presenting a cohesive exploration of computational drug design strategies applied across diverse and mechanistically complex disease domains.

This book presents molecular dynamics simulations, which establish the physicochemical foundation of structure-informed therapeutic modeling. Atomistic trajectory analysis, free-energy estimation, enhanced sampling algorithms, and machine learning-augmented simulation platforms provide quantitative insight into conformational plasticity, allosteric modulation, and ligand–target stability. These methodologies transform static structural representations into dynamic mechanistic models capable of resolving thermodynamic and kinetic determinants of binding.

This structural perspective is complemented by rigorous examination of in silico ADMET prediction. Pharmacokinetic behavior, metabolic liability, cytochrome P450 modulation, transporter interactions, organ-specific toxicity, and clearance parameters are evaluated through quantitative structure–property relationships and advanced artificial intelligence architectures. Together, these analytical layers reinforce a central principle of modern therapeutic design: binding affinity alone is insufficient. Complex disease biology necessitates multidimensional therapeutic strategies. Multi-target and polypharmacology-oriented design approaches articulate a systems-level paradigm in which distributed modulation of interconnected pathways supersedes reductionist single-node inhibition. Such conceptual realignment is particularly relevant in oncology, where signaling redundancy, pathway cross-talk, and tumor heterogeneity demand computational pipelines that integrate structure-based modeling, virtual screening, network pharmacology, and algorithm-guided optimization to achieve durable therapeutic response.

Structure-informed methodologies extend with equal significance to infectious disease therapeutics, where atomic-resolution characterization of viral enzymes, bacterial biosynthetic machinery, and host–pathogen interfaces enables precise vulnerability mapping. Parallel computational frameworks address cardiometabolic disorders, modeling convergent dysregulation of glucose metabolism, lipid homeostasis, oxidative stress, inflammatory signaling, and receptor-mediated control systems. Neurodegenerative conditions introduce further complexity through protein misfolding kinetics, mitochondrial dysfunction, neuroinflammatory cascades, and central nervous system delivery constraints, underscoring the importance of systems biology integration, structural annotation, and blood–brain barrier permeability prediction in target prioritization.

Across these therapeutic domains, a consistent scientific principle becomes evident: progress is increasingly driven by integrated computational frameworks rather than isolated methodological advances. Molecular docking, molecular dynamics simulations, pharmacophore construction, quantitative structure–activity relationships, graph-based machine learning models, deep neural network architectures, and systems-level target prioritization function as coordinated elements within iterative and feedback-driven discovery pipelines. Their collective value lies in methodological complementarity and cross-validation rather than a standalone application.

Equally essential is disciplined acknowledgment of methodological limitations. Algorithm interpretability, dataset representativeness, domain boundaries, reproducibility, translational uncertainty, and regulatory oversight of artificial intelligence-assisted decision systems demand sustained analytical vigilance. Computational refinement cannot substitute for biological validation. Predictive models attain scientific credibility only when supported by experimental evidence and aligned with established mechanistic principles. This volume is intended for researchers in computational chemistry, structural biology, pharmacology, medicinal chemistry, systems biology, and translational therapeutics. It provides theoretical foundations alongside application-oriented analyses. Advances in high-performance computing, multi-omics integration, federated data infrastructures, and explainable artificial intelligence continue to redefine therapeutic design. As molecular mechanisms of disease are resolved with increasing precision, computational strategy is not going to merely accelerate discovery but reshape its scientific architecture.

About the editors

Mohammad Kalim Ahmad Khan is Professor of Bioengineering and Course Coordinator for M.Tech (Bioinformatics) and B.Tech (Biotechnology) at Integral University, Lucknow. An M.Sc. alumnus of the same institution, he has cultivated more than 16 years of expertise in computer-aided drug design, high-throughput virtual screening, ADMET profiling, and molecular dynamics simulations. He has authored 82 peer-reviewed papers, two research monographs, and three book chapters. He is also an invited reviewer for over 30 international journals. His laboratory focuses on structure-based inhibitor design for breast cancer targets, emerging tuberculosis enzymes, and chemokine-mediated vitiligo pathways. In recognition of his scholarly contributions, he received the 2022 International Eminent Award in Engineering, Science, and Medicine. Within the university, he leads National Assessment and Accreditation Council (NAAC) quality assurance initiatives, manages the Postgraduate Computational Biology Laboratory, and serves on the Board of Studies, Faculty Board, and Academic Council. He is a member of IAENG, IACSIT, and ISRD.

Mohammed Tarique is a founder and Director of Alamanc Life Sciences, New Delhi, India. Previously, he has worked at the Centre for Interdisciplinary Research in Basic Sciences (CIRBSc) at Jamia Millia Islamia, India. Holding a Ph.D. in Plasmodium Biology from the International Centre for Genetic Engineering and Biotechnology, New Delhi, India, his research focuses on understanding the biology of *Plasmodium* for the development of novel drugs. He has published more than 40 research articles in peer-reviewed international journals. He currently serves on the editorial boards of numerous journals and has published several books with international publishers.

Chapter 1

Molecular dynamics simulations in drug discovery

Integration with artificial intelligence and future viewpoints

Sangavi Pandi[1,2], *Dapkupar Wankhar*[3], *J. H. Shazia Fathima*[4], *Priyanka Ganapathy*[5], *Langeswaran Kulanthaivel*[6], *and Gowtham Kumar Subbaraj*[7]

[1]Center for Bioinformatics, Karpagam Academy of Higher Education, Coimbatore, Tamil Nadu, India

[2]Department of Microbiology, Karpagam Academy of Higher Education, Coimbatore, Tamil Nadu, India

[3]Faculty of Allied and Healthcare Sciences, Assam Down Town University, Sankar Madhab Path, Gandhi Nagar, Guwahati, Assam, India

[4]Global Research Cell, Dr. D. Y. Patil Dental College & Hospital, Dr. D. Y. Patil Vidyapeeth (Deemed to be University), Pimpri, Pune, India

[5]Department of Physiology, Sree Balaji Medical College and Hospital, Chennai, Tamilnadu, India

[6]Department of Biomedical Science, Alagappa University, Karaikudi, Tamil Nadu, India

[7]Saveetha Institute of Basic Medical Sciences (SIBMS), Saveetha Institute of Medical and Technical Sciences (SIMATS), Saveetha University, Thandalam, Chennai, Tamil Nadu, India

1.1 INTRODUCTION

The process of drug discovery is inherently challenging and resource-intensive, necessitating a precise understanding of the molecular interactions that underpin biological activity. A critical facet of this process involves characterizing the interactions between small molecules and biological macromolecules, such as proteins, nucleic acids, and membranes [1]. Experimental structural biology techniques, including X-ray crystallography, nuclear magnetic resonance (NMR) spectroscopy, and cryo-electron microscopy

DOI: 10.1201/9781003672760-1

(cryo-EM), have provided invaluable structural information regarding drug targets. However, these methodologies predominantly yield static representations of biomolecules, which are often captured under non-physiological conditions. In biological systems, macromolecules are intrinsically dynamic [2]. Proteins undergo continuous conformational fluctuations that are essential for ligand recognition, enzymatic catalysis, signal transduction, and regulatory functions. Many pharmacologically relevant phenomena, such as induced fit, conformational selection, allosteric modulation, and the formation of cryptic pockets, cannot be fully comprehended through static structures alone. As a result, computational approaches capable of capturing molecular motion have become indispensable within modern drug discovery [3]. Recent advancements in computer-based analysis have made it possible to screen drug candidates more effectively from target proteins and chemical libraries. However, predictions made using artificial intelligence (AI) and simulation technologies still require experimental confirmation, as they are crucial for narrowing down a vast number of possibilities. The accuracy and logic of these predictions are essential, especially since experiments can be very costly.

Computer-Aided Drug Design (CADD) is a specialized field that simulates the interactions between drugs and receptors using computational techniques. CADD approaches can be divided into two main categories: Ligand-Based Drug Design (LBDD) and Structure-Based Drug Design (SBDD) [4]. SBDD requires knowledge of the three-dimensional structures of target proteins, while LBDD can be employed when these structures are not available. Thanks to advancements in structural biology, such as cryo-EM, and computational protein structure prediction methods like AlphaFold, there is now a greater availability of high-resolution 3D structures of new drug-receptor complexes. This convergence of discoveries has transformed SBDD into a powerful computational technique that facilitates the development of new small-molecule treatments [5].

Molecular dynamics (MD) simulations offer a physics-based framework for examining the time-dependent behavior of molecular systems by explicitly modeling atomic motions under defined thermodynamic conditions. By solving the equations of motion for all atoms within a system, MD simulations generate trajectories that elucidate how biomolecules evolve. These trajectories provide insights into structural stability, flexibility, intermolecular interactions, and energetic landscapes that would otherwise remain inaccessible [6]. With the rapid advancement of high-performance computing and the development of robust simulation software, MD simulations have transitioned from specialized academic tools to routine components of industrial drug discovery pipelines. Presently, MD simulations are widely employed to refine docking poses, estimate binding free energies, investigate resistance mechanisms, characterize allosteric sites, and guide rational lead optimization. This chapter provides a comprehensive overview of MD simulations, focusing on both the methodological aspects and practical applications relevant to drug discovery.

1.2 FUNDAMENTAL AND ADVANCED MOLECULAR DYNAMICS SIMULATION METHODS

1.2.1 Classical molecular dynamics simulations

Classical MD simulations constitute the most prevalent and foundational methodology in biomolecular modeling and drug discovery. In classical MD, atoms are modeled as point masses, and their movements are described by Newton's equations of motion, utilizing predefined force fields. This methodology facilitates the explicit modeling of atomic interactions over time, thereby enabling the exploration of structural stability, conformational flexibility, and intermolecular interactions under conditions that closely resemble physiological environments [7,8]. In the realm of drug discovery, classical MD simulations are extensively employed to assess the stability of protein–ligand complexes after molecular docking has occurred. By monitoring parameters such as root mean square deviation (RMSD), hydrogen bond occupancy, and interaction energies, researchers can evaluate the persistence and quality of binding interactions [9]. Additionally, classical MD provides critical insights into induced-fit effects, solvent-mediated interactions, and side-chain rearrangements that significantly influence ligand affinity. However, a notable limitation of classical MD is its constrained ability to sample events over extended timescales, with simulations typically confined to durations ranging from nanoseconds to microseconds [10].

1.2.2 Coarse-grained molecular dynamics simulations

Coarse-grained (CG) MD simulations simplify system complexity by aggregating multiple atoms into singular interaction sites, commonly referred to as "beads." This methodological simplification significantly reduces the particle count within the system, thereby facilitating the exploration of extended timescales and larger spatial dimensions compared to traditional atomistic MD simulations [11,12]. CG models are particularly advantageous for investigating large biomolecular assemblies, such as membranes, viral capsids, and multi-protein complexes [13]. Within the framework of drug discovery, CG MD simulations are routinely employed to examine large-scale conformational changes, protein–protein interactions, and membrane-associated targets, including G protein-coupled receptors. Although CG simulations relinquish atomic-level detail, they provide essential insights into global dynamics and long-range interactions that remain largely inaccessible via standard MD approaches [14]. An increasing trend toward hybrid methodologies, which integrate CG and atomistic representations, has emerged to achieve a balance between computational efficiency and structural accuracy.

1.2.3 Steered molecular dynamics simulations

Steered molecular dynamics (SMD) simulations represent a nonequilibrium MD technique wherein external forces are applied to specific atoms or molecular regions to induce conformational transitions [15]. By exerting a controlled force along a predetermined reaction coordinate, SMD simulations can effectively mimic experimental techniques such as atomic force microscopy and optical tweezers [16]. In the domain of drug discovery, SMD is frequently utilized to investigate ligand unbinding pathways, estimate relative binding affinities, and identify critical residues that play significant roles in dissociation processes. The analysis of the force–distance profiles generated during SMD simulations yields qualitative and quantitative insights into interaction strengths and energy barriers [17]. However, careful attention must be paid to the selection of pulling velocities and force constants to avoid potential artifacts that may arise from nonequilibrium effects, complicating the interpretation of the results.

1.2.4 Accelerated molecular dynamics simulations

Accelerated molecular dynamics (aMD) constitutes an advanced sampling technique designed to mitigate the timescale limitations inherent in conventional MD simulations [18]. In aMD, a biasing potential is introduced to the system's potential energy surface, effectively lowering energy barriers and allowing for more frequent transitions between conformational states. This enhancement facilitates the exploration of infrequently visited yet biologically relevant conformations within a practical computational timeframe [19]. Accelerated MD has demonstrated considerable utility in drug discovery, particularly for the identification of allosteric binding sites, the examination of conformational flexibility, and the characterization of activation or inactivation mechanisms of drug targets. By producing broader conformational ensembles, aMD contributes to an improved understanding of ligand accessibility and binding adaptability. Nonetheless, reweighting procedures are often necessary to accurately recover thermodynamic properties, thereby adding complexity to subsequent data analysis [20].

1.2.5 Replica exchange molecular dynamics simulations

Replica exchange molecular dynamics (REMD) is an effective enhanced sampling strategy that involves the concurrent simulation of multiple replicas of a system at varying temperatures or Hamiltonians [21].

Table 1.1 Overview of molecular dynamics simulation methods used in drug discovery

MD method	*Description*	*Accessible timescale*	*Key advantages*	*Limitations*	*Typical applications*
Classical MD	Atomistic simulation based on Newtonian mechanics	ns–μs	High structural detail; widely validated	Limited sampling of rare events	Protein–ligand stability, conformational analysis
Coarse-grained MD	Group atoms into pseudo-particles	μs–ms	Faster simulations; large systems	Reduced atomic detail	Membrane proteins, large complexes
Steered MD	External force applied to induce transitions	ns–μs	Studies unbinding pathways	Nonequilibrium effects	Ligand dissociation studies
Accelerated MD	Boosts the potential energy surface	μs–ms	Enhanced sampling	Requires careful parameter tuning	Allosteric site exploration
Replica Exchange MD	Multiple replicas at different temperatures	μs	Efficient barrier crossing	High computational cost	Protein folding, conformational sampling

Periodic exchanges between these replicas facilitate the efficient crossing of energy barriers, resulting in enhanced sampling of the conformational space. In the context of drug discovery, REMD is particularly valuable for analyzing protein folding, conformational heterogeneity, and the flexible binding sites that influence ligand recognition [22]. This method yields well-equilibrated ensembles that can be employed for docking, free energy calculations, and pharmacophore modeling. Despite its benefits, REMD is computationally demanding due to the necessity of simulating multiple replicas in parallel, which may limit its routine application to smaller systems or specific targeted investigations [23] (Table 1.1).

1.3 THEORETICAL FOUNDATIONS OF MOLECULAR DYNAMICS SIMULATIONS

1.3.1 Classical mechanics and atomic motion

MD simulations are primarily based on classical mechanics, where atoms are regarded as point masses that interact according to predefined potential energy functions [24]. The motion of each atom is governed by Newton's second law of motion, which establishes a relationship between the force acting upon an atom, its mass, and its acceleration [25]. Forces are obtained as the negative gradient of the potential energy function with respect to atomic coordinates. The equations of motion are integrated numerically using finite time steps, typically on the order of femtoseconds. Given that biomolecular systems may contain thousands to millions of atoms, analytical solutions are generally impractical, thereby necessitating the use of numerical integration schemes. Commonly employed algorithms include Verlet, leapfrog, and velocity-Verlet, which are favored for their stability and efficiency. While MD simulations are grounded in classical mechanics, they provide a satisfactory approximation for most biological processes relevant to drug discovery, as quantum effects are typically confined to specific regions, such as enzymatic active sites.

1.3.2 Potential energy functions and force fields

The accuracy of MD simulations is critically contingent upon the quality of the force field, which delineates the mathematical form and parameters of the potential energy function. Force fields disaggregate the total potential energy into bonded and non-bonded components [26]. The bonded terms encapsulate covalent interactions, including bond stretching, angle bending, and dihedral torsions, whereas the non-bonded terms account for electrostatic interactions and van der Waals forces. Several force fields have been established for biomolecular simulations, including AMBER [27], CHARMM [28], OPLS-AA [29], and GROMOS [30]. These force fields differ in their parameterization methodologies and are subject to ongoing refinement to enhance accuracy. In the context of drug discovery, particular attention must be devoted to ligand parameterization, as small molecules frequently exhibit chemical features that are not represented in conventional biomolecular force fields.

1.3.3 Statistical mechanics and thermodynamic ensembles

MD simulations are firmly anchored in statistical mechanics, which offers the theoretical framework that connects microscopic atomic movements to macroscopic thermodynamic properties [31]. Simulations are conducted within specific ensembles, such as NVT (constant number of particles, volume, and temperature) or NPT (constant number of particles, pressure, and temperature) [32]. Thermostats and barostats are employed to maintain the desired temperature and pressure conditions, thereby ensuring appropriate sampling of the phase space [33] (Table 1.2).

Table 1.2 Common force fields used in molecular dynamics simulations for drug discovery

Force field	*System type*	*Key features*	*Strengths*	*Limitations*	*Typical usage*
AMBER	Proteins, nucleic acids, ligands	Fixed-charge, biomolecule-optimized	High accuracy for proteins	Ligand parameterization required	Enzyme–ligand studies
CHARMM	Proteins, lipids, membranes	Extensive lipid parameters	Excellent membrane modeling	Computationally intensive	GPCR and membrane targets
OPLS-AA	Proteins and small molecules	Balanced non-bonded interactions	Good ligand representation	Limited nucleic acid support	Drug–target binding studies
GROMOS	Biomolecules	United-atom approach	Fast simulations	Lower resolution	Long-time scale dynamics

1.4 SYSTEM PREPARATION FOR DRUG DISCOVERY SIMULATIONS

1.4.1 Protein structure preparation

Protein preparation represents an essential step in MD simulations [34]. Experimental structures may contain missing residues, unresolved loops, alternate conformations, or crystallographic artifacts. These challenges necessitate meticulous preprocessing, including model building, loop refinement, assignment of protonation states, and optimization of hydrogen bonding networks [35]. The accurate assignment of protonation states for ionizable residues is particularly critical, as these states significantly influence electrostatic interactions and ligand binding. Typically, the assignment of protonation states is based on physiological pH and the local microenvironment [36].

1.4.2 Ligand preparation and parameterization

Ligand preparation entails the generation of precise three-dimensional geometries, the assignment of partial charges, and the parameterization of the molecule using compatible force field parameters [37]. Inaccuracies in ligand parameterization can significantly impact simulation outcomes, especially in binding free energy calculations. Quantum mechanical calculations are frequently employed to derive high-quality charges for ligands, ensuring the reliability of the simulations [38].

1.4.3 Solvation and ionic environment

To accurately represent biological conditions, systems are solvated using explicit water models. Counterions are introduced to neutralize the system and replicate physiological ionic strength [39]. Solvent molecules are pivotal in mediating protein–ligand interactions, stabilizing conformations, and influencing the thermodynamics of binding [40].

1.4.4 Energy minimization and equilibration

Energy minimization is a critical step in MD simulations, aimed at resolving any steric clashes and alleviating high-energy contacts that might have been introduced during the preparatory phase of the system [41]. This process involves adjusting the positions of atoms to achieve a more stable conformation. Following energy minimization, the system undergoes a series of equilibration phases. During these phases, the system is systematically heated to a target temperature and equilibrated under controlled pressure conditions [42]. This gradual acclimatization process is essential for ensuring that the molecular system reaches a realistic and physically meaningful state before any data collection begins, thus facilitating accurate and reliable results.

1.4.5 Production molecular dynamics simulations

Production MD simulations are meticulously conducted to investigate the conformational landscape of the biomolecular system being studied [43]. The duration of these simulations can impact the insights gained and varies widely depending on the specific biological question or hypothesis in focus. For instance, simulations lasting around tens of nanoseconds may suffice for preliminary stability assessments, while studies investigating large-scale conformational changes often require simulations extending into microseconds or longer [44]. The trajectories generated during this production phase capture detailed atomic positions and velocities over time, ultimately forming the basis for a wide array of analytical approaches that reveal vital biochemical properties and states [45].

1.4.6 Trajectory analysis and interpretation

Trajectory analysis is an indispensable component of understanding the results obtained from MD simulations [46]. Studies employ various metrics to evaluate structural stability, including RMSD to track changes in overall structure, root mean square fluctuation (RMSF) to assess the flexibility of particular residues, and the radius of gyration, which provides insight into the compactness of the molecular structure [47]. Interaction analyses delve deeper, focusing on specific interactions such as hydrogen bonds, salt bridges, π–π stacking interactions, and hydrophobic contacts, all of which are pivotal in understanding molecular behavior [48]. Additionally, advanced analytical techniques such as principal component analysis (PCA) facilitate the exploration of collective motions, while dynamic cross-correlation

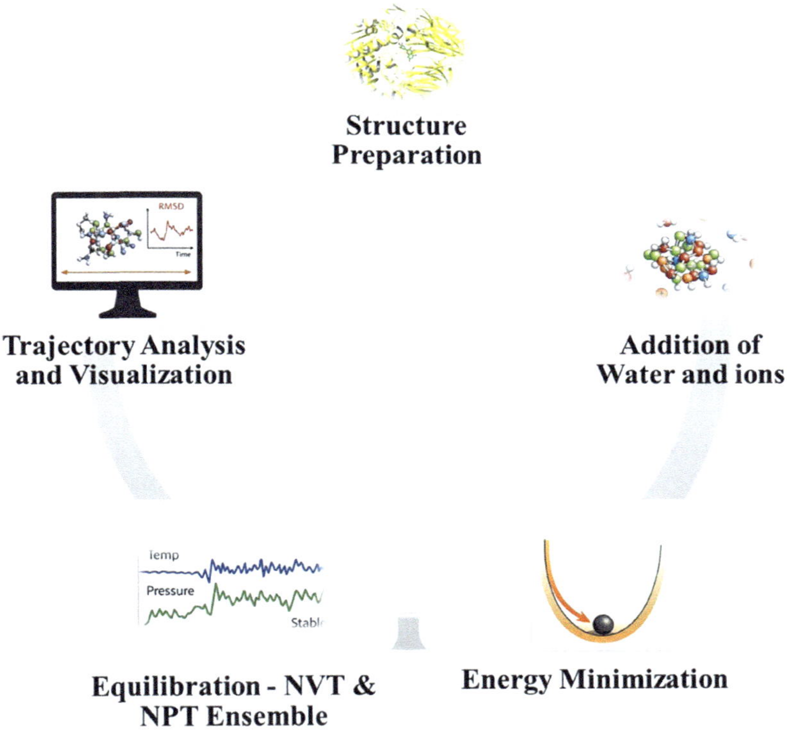

Figure 1.1 Schematic representation of key steps involved in performing molecular dynamics simulation.

matrices reveal how different regions of the molecule move in relation to one another [49,50] Furthermore, constructing free energy landscapes aids in visualizing the conformational transitions that are particularly relevant for understanding ligand binding and functional mechanisms (Figure 1.1).

1.5 ENHANCED SAMPLING METHODS

Traditional MD simulations often struggle with the challenge of efficiently sampling rare events or overcoming significant energy barriers that may exist within the system. To address these limitations, researchers have developed enhanced sampling techniques [51]. Methods such as metadynamics introduce a bias to the energy landscape to facilitate transitions between states, while accelerated MD accelerates the dynamics of the system to explore its conformational space more rapidly [52]. Replica exchange MD allows multiple simulations at different temperatures to share information about configuration space, and umbrella sampling employs restraint forces to sample regions of interest systematically. These enhanced sampling methodologies significantly improve simulation efficiency, enabling a more thorough exploration of complex energy landscapes and dynamic processes [53].

1.6 FREE ENERGY CALCULATIONS IN DRUG DISCOVERY

One of the most impactful applications of MD simulations in drug discovery is the estimation of binding free energy, which serves as a crucial metric for predicting the affinity of potential drug candidates for their target proteins [54]. Various methods are utilized for this purpose, with end-point approaches like molecular mechanics with Poisson–Boltzmann surface area (MM/PBSA) and molecular mechanics with generalized Born surface area (MM/GBSA) offering relatively fast yet effective estimates for ranking ligands based on binding affinity [55]. In contrast, alchemical methods, including free energy perturbation and thermodynamic integration, provide more accurate calculations of binding free energies and are increasingly favored during the lead optimization phase of drug design [56]. These calculations are vital for identifying promising drug candidates that exhibit strong interactions with their biological targets.

1.7 APPLICATIONS OF MOLECULAR DYNAMICS SIMULATIONS IN DRUG DISCOVERY

MD simulations are instrumental at multiple stages throughout the drug discovery process. They are used to validate docking poses from computational ligand-screening methods and to refine binding modes to improve the accuracy of predicted interactions [57]. Furthermore, MD simulations aid in guiding the optimization of lead compounds by providing insights into modifications that enhance binding affinity [58]. MD simulations elucidate bioactive–protein interactions, genetic variant effects, and network-derived targets, accelerating AI-assisted cardiovascular drug discovery and therapeutic optimization [59–63]. These simulations also prove essential for identifying allosteric sites that may offer novel therapeutic targets, as well as for investigating mechanisms of resistance that pathogens may develop against existing drugs. Beyond these applications, MD simulations enrich our understanding of elementary processes, such as enzyme catalysis, receptor activation, and signal transduction pathways, thereby providing deeper mechanistic insight into potential drug actions.

1.8 INTEGRATION OF MOLECULAR DYNAMICS WITH ARTIFICIAL INTELLIGENCE

The integration of MD simulations with AI and machine learning has opened up innovative avenues for accelerating the drug discovery process [64]. By employing advanced machine learning algorithms, researchers can analyze extensive MD datasets to identify patterns and correlations that may be missed through traditional approaches. These models can also be used to refine force field parameters, resulting in more accurate simulations [65]. Adaptive sampling strategies guided by machine learning allow researchers to focus on the most relevant parts of the conformational space, enhancing the overall efficiency of the drug discovery process [66]. This synergy between AI and MD simulations holds great promise for revolutionizing how drugs are developed and tailored to address complex biological challenges.

1.9 LIMITATIONS AND CHALLENGES

Despite the considerable advantages offered by MD simulations, several limitations remain, particularly concerning computational cost, sampling efficiency, and the accuracy of force fields. The high resource requirements for running extensive simulations can be a significant barrier, especially when addressing large systems or long time scales. Furthermore, poor sampling can lead to incomplete exploration of conformational states, resulting in the misinterpretation of dynamics. Finally, the accuracy of the force fields employed, which are essential for simulating molecular interactions, can vary, leading to discrepancies between simulation results and experimental data. It is thus crucial to rigorously validate computational findings against experimental observations and maintain a critical perspective when interpreting results.

1.10 FUTURE PERSPECTIVES

Looking ahead, advances in high-performance computing, cloud-based platforms, and the continued integration of AI-assisted methodologies are poised to expand the capabilities of MD simulations within drug discovery. As computational technologies mature and become more accessible, MD simulations will increasingly contribute to predictive and personalized therapeutic design, empowering researchers to tailor drug candidates to individual patient profiles and specific diseases. These developments hold the promise of enhancing the speed and precision of drug discovery efforts, ultimately leading to more effective treatments.

1.11 CONCLUSION

In summary, MD simulations have emerged as a pivotal and indispensable tool in contemporary drug discovery. They allow researchers to explore biomolecular systems dynamically, revealing time-dependent behaviors that static structural representations cannot capture. By providing critical insights into conformational flexibility, binding stability, and the intricate molecular mechanisms that underpin drug action, MD simulations significantly enhance the efficiency and effectiveness of drug design processes. The integration of MD simulations with structure-based approaches has dramatically improved hit validation, lead optimization, and the identification of allosteric binding sites. Thanks to advanced techniques such as enhanced sampling methods and binding free energy calculations, MD simulations are now integral in addressing complex biological questions. While challenges related to computational resources and force field accuracy persist, ongoing methodological advancements continue to enhance the reliability and efficiency. The increasing collaboration between MD simulations and AI, paired with breakthroughs in high-performance computing, promises to further accelerate drug discovery initiatives. Consequently, MD simulations will remain a central component of modern computational drug discovery and play a critical role in rational therapeutic design in the future.

REFERENCES

1. S.K. Niazi, Z. Mariam, Computer-Aided Drug Design and Drug Discovery: A Prospective Analysis, *Pharmaceuticals* 17 (2023) 22. https://doi.org/10.3390/ph17010022.
2. S.-F. Zhou, W.-Z. Zhong, Drug Design and Discovery: Principles and Applications, *Molecules* 22 (2017) 279. https://doi.org/10.3390/molecules22020279.
3. N. Singh, P. Vayer, S. Tanwar, J.-L. Poyet, K. Tsaioun, B.O. Villoutreix, Drug Discovery and Development: Introduction to the General Public and Patient Groups, *Frontiers in Drug Discovery* 3 (2023) 1201419. https://doi.org/10.3389/fddsv.2023.1201419.
4. J.-P. Renaud, A. Chari, C. Ciferri, W. Liu, H.-W. Rémigy, H. Stark, C. Wiesmann, Cryo-EM in Drug Discovery: Achievements, Limitations and Prospects, *Nature Reviews Drug Discovery* 17 (2018) 471–492. https://doi.org/10.1038/nrd.2018.77.
5. T.I. Adelusi, A.-Q.K. Oyedele, I.D. Boyenle, A.T. Ogunlana, R.O. Adeyemi, C.D. Ukachi, M.O. Idris, O.T. Olaoba, I.O. Adedotun, O.E. Kolawole, Y. Xiaoxing, M. Abdul-Hammed, Molecular Modeling in Drug Discovery, *Informatics in Medicine Unlocked* 29 (2022) 100880. https://doi.org/10.1016/j.imu.2022.100880.

6. H. Iwata, Transforming Drug Discovery: The Impact of AI and Molecular Simulation on R&D Efficiency, *Bioanalysis* 16 (2024) 1211–1217. https://doi.org/10.1080/17576180.2024.2437283.
7. S.A. Hollingsworth, R.O. Dror, Molecular Dynamics Simulation for All, *Neuron* 99 (2018) 1129–1143. https://doi.org/10.1016/j.neuron.2018.08.011.
8. H.M. Senn, W. Thiel, QM/MM Methods for Biomolecular Systems, *Angewandte Chemie International Edition* 48 (2009) 1198–1229. https://doi.org/10.1002/anie.200802019.
9. L. Ferreira, R. Dos Santos, G. Oliva, A. Andricopulo, Molecular Docking and Structure-Based Drug Design Strategies, *Molecules* 20 (2015) 13384–13421. https://doi.org/10.3390/molecules200713384.
10. O.M.H. Salo-Ahen, I. Alanko, R. Bhadane, A.M.J.J. Bonvin, R.V. Honorato, S. Hossain, A.H. Juffer, A. Kabedev, M. Lahtela-Kakkonen, A.S. Larsen, E. Lescrinier, P. Marimuthu, M.U. Mirza, G. Mustafa, A. Nunes-Alves, T. Pantsar, A. Saadabadi, K. Singaravelu, M. Vanmeert, Molecular Dynamics Simulations in Drug Discovery and Pharmaceutical Development, *Processes* 9 (2020) 71. https://doi.org/10.3390/pr9010071.
11. C. Giavarini, D. Mastrofini, M. Scarsella, L. Barré, D. Espinat, Macrostructure and Rheological Properties of Chemically Modified Residues and Bitumens, *Energy & Fuels* 14 (2000) 495–502. https://doi.org/10.1021/ef9902045.
12. E.I. Assaf, X. Liu, P. Lin, S. Erkens, Introducing a Force-Matched United Atom Force Field to Explore Larger Spatiotemporal Domains in Molecular Dynamics Simulations of Bitumen, *Materials & Design* 240 (2024) 112831. https://doi.org/10.1016/j.matdes.2024.112831.
13. Z. Zhang, J. Pfaendtner, A. Grafmüller, G.A. Voth, Defining Coarse-Grained Representations of Large Biomolecules and Biomolecular Complexes from Elastic Network Models, *Biophysical Journal* 97 (2009) 2327–2337. https://doi.org/10.1016/j.bpj.2009.08.007.
14. H.J. Kim, J.-S. Kim, Y. Kim, Y.S. Jung, B.J. Kim, Y. Kim, Regioregularity Controlled Phase Behavior for Poly(3-Hexylthiophene): A Combined Study of Simple Coarse-Grained Simulation and Experiment, *Polymer (Guildf)* 178 (2019) 121569. https://doi.org/10.1016/j.polymer.2019.121569.
15. Y. Yu, J. Weng, W. Wang, Use of Steered Molecular Dynamics to Explore the Conformational Changes of SNARE Proteins, in: Fratti, R. (eds), *SNAREs. Methods in Molecular Biology*. Humana, New York, 2025: pp. 69–77. https://doi.org/10.1007/978-1-0716-4314-3_4.
16. R. Tomarchio, V. Patamia, C. Zagni, L. Crocetti, A. Cilibrizzi, G. Floresta, A. Rescifina, Steered Molecular Dynamics Simulations Study on FABP4 Inhibitors, *Molecules* 28 (2023) 2731. https://doi.org/10.3390/molecules28062731.
17. M. Suan Li, B. Khanh Mai, Steered Molecular Dynamics-A Promising Tool for Drug Design, *Current Bioinformatics* 7 (2012) 342–351. https://doi.org/10.2174/157489312803901009.
18. Y. Wang, C.B. Harrison, K. Schulten, J.A. McCammon, Implementation of Accelerated Molecular Dynamics in NAMD, *Computational Science & Discovery* 4 (2011) 015002. https://doi.org/10.1088/1749-4699/4/1/015002.

19. Y.T. Pang, Y. Miao, Y. Wang, J.A. McCammon, Gaussian Accelerated Molecular Dynamics in NAMD, *Journal of Chemical Theory and Computation* 13 (2017) 9–19. https://doi.org/10.1021/acs.jctc.6b00931.
20. Y. Miao, W. Sinko, L. Pierce, D. Bucher, R.C. Walker, J.A. McCammon, Improved Reweighting of Accelerated Molecular Dynamics Simulations for Free Energy Calculation, *Journal of Chemical Theory and Computation* 10 (2014) 2677–2689. https://doi.org/10.1021/ct500090q.
21. C. Bergonzo, N.M. Henriksen, D.R. Roe, J.M. Swails, A.E. Roitberg, T.E. Cheatham, Multidimensional Replica Exchange Molecular Dynamics Yields a Converged Ensemble of an RNA Tetranucleotide, *Journal of Chemical Theory and Computation* 10 (2014) 492–499. https://doi.org/10.1021/ct400862k.
22. R. Lazim, D. Suh, S. Choi, Advances in Molecular Dynamics Simulations and Enhanced Sampling Methods for the Study of Protein Systems, *International Journal of Molecular Sciences* 21 (2020) 6339. https://doi.org/10.3390/ijms21176339.
23. A. Gil-Ley, G. Bussi, Enhanced Conformational Sampling Using Replica Exchange with Collective-Variable Tempering, *Journal of Chemical Theory and Computation* 11 (2015) 1077–1085. https://doi.org/10.1021/ct5009087.
24. G. Ciccotti, S. Decherchi, S. Meloni, Foundations of Molecular Dynamics Simulations: How and What, *La Rivista Del Nuovo Cimento* 48 (2025) 1–94. https://doi.org/10.1007/s40766-025-00065-4.
25. A.P. Thompson, H.M. Aktulga, R. Berger, D.S. Bolintineanu, W.M. Brown, P.S. Crozier, P.J. in 't Veld, A. Kohlmeyer, S.G. Moore, T.D. Nguyen, R. Shan, M.J. Stevens, J. Tranchida, C. Trott, S.J. Plimpton, LAMMPS—A Flexible Simulation Tool for Particle-Based Materials Modeling at the Atomic, Meso, and Continuum Scales, *Computer Physics Communications* 271 (2022) 108171. https://doi.org/10.1016/j.cpc.2021.108171.
26. S. Kashefolgheta, S. Wang, W.E. Acree, P.H. Hünenberger, Evaluation of Nine Condensed-Phase Force Fields of the GROMOS, CHARMM, OPLS, AMBER, and OpenFF Families Against Experimental Cross-Solvation Free Energies, *Physical Chemistry Chemical Physics* 23 (2021) 13055–13074. https://doi.org/10.1039/D1CP00215E.
27. J.W. Ponder, D.A. Case, Force Fields for Protein Simulations, *Advances in Protein Chemistry* 66 (2003) 27–85. https://doi.org/10.1016/S0065-3233(03)66002-X.
28. B.R. Brooks, C.L. Brooks, A.D. Mackerell, L. Nilsson, R.J. Petrella, B. Roux, Y. Won, G. Archontis, C. Bartels, S. Boresch, A. Caflisch, L. Caves, Q. Cui, A.R. Dinner, M. Feig, S. Fischer, J. Gao, M. Hodoscek, W. Im, K. Kuczera, T. Lazaridis, J. Ma, V. Ovchinnikov, E. Paci, R.W. Pastor, C.B. Post, J.Z. Pu, M. Schaefer, B. Tidor, R.M. Venable, H.L. Woodcock, X. Wu, W. Yang, D.M. York, M. Karplus, CHARMM: The Biomolecular Simulation Program, *Journal of Computational Chemistry* 30 (2009) 1545–1614. https://doi.org/10.1002/jcc.21287.

29. W.L. Jorgensen, D.S. Maxwell, J. Tirado-Rives, Development and Testing of the OPLS All-Atom Force Field on Conformational Energetics and Properties of Organic Liquids, *Journal of the American Chemical Society* 118 (1996) 11225–11236. https://doi.org/10.1021/ja9621760.
30. M.M. Reif, M. Winger, C. Oostenbrink, Testing of the GROMOS Force-Field Parameter Set 54A8: Structural Properties of Electrolyte Solutions, Lipid Bilayers, and Proteins, *Journal of Chemical Theory and Computation* 9 (2013) 1247–1264. https://doi.org/10.1021/ct300874c.
31. Q. Mao, M. Feng, X.Z. Jiang, Y. Ren, K.H. Luo, A.C.T. van Duin, Classical and Reactive Molecular Dynamics: Principles and Applications in Combustion and Energy Systems, *Progress in Energy and Combustion Science* 97 (2023) 101084. https://doi.org/10.1016/j.pecs.2023.101084.
32. P. Sangavi, H. Nagarajan, S. Subramaniyan, J. Jeyaraman, K. Langeswaran, Unveiling the Oncological Inhibition of Bioactive Compounds from *Adansonia digitata* via *in silico* Analysis by Targeting γ-Butyrobetaine Dioxygenase 1 Against Triple Negative Breast Cancer, *Journal of Biomolecular Structure and Dynamics* 44 (2026) 234–257. https://doi.org/10.1080/07391102.2024.2437528.
33. Q. Ke, X. Gong, S. Liao, C. Duan, L. Li, Effects of Thermostats/Barostats on Physical Properties of Liquids by Molecular Dynamics Simulations, *Journal of Molecular Liquids* 365 (2022) 120116. https://doi.org/10.1016/j.molliq.2022.120116.
34. S. Pandi, N. Kathiresan, G. Kumar Subbaraj, D. Desai, C. Nagarajan, L. Kulanthaivel, Decoding the Anticancer and Biofilm-Inhibiting Efficacy of Adansonia Digitata Using Experimental, AI-Powered, and Molecular Modeling Approaches, *Frontiers in Molecular Biosciences* 12 (2025) 1666360. https://doi.org/10.3389/fmolb.2025.1666360.
35. L. Heo, M. Feig, Experimental Accuracy in Protein Structure Refinement via Molecular Dynamics Simulations, *Proceedings of the National Academy of Sciences* 115 (2018) 13276–13281. https://doi.org/10.1073/pnas.1811364115.
36. A. Prajapati, R. Kachhadiya, A. Pravin, S.K. Singh, H. Patel, S. Patel, Computational Identification of Marine Natural Products as Potential MDM2 Inhibitors Using E-Pharmacophore-Based Virtual Screening, MD Simulations, MM-GBSA, DFT, PCA, and FEL Analysis, *ChemistrySelect* 10 (2025) e01564. https://doi.org/10.1002/slct.202501564.
37. P. Sangavi, G.R. Shri, S.K. Singh, K. Langeswaran, A Computational Perception of BBOX1-IP3R3 Interaction Uncovers Inhibitors for Dysregulated Calcium Signalling in Triple Negative Breast Cancer, *SAR and QSAR in Environmental Research* 36 (2025) 305–332. https://doi.org/10.1080/1062936X.2025.2497380.
38. M. Osato, H.M. Baumann, J. Huang, I. Alibay, D.L. Mobley, Evaluating the Functional Importance of Conformer-Dependent Atomic Partial Charge Assignment, *Journal of Computational Chemistry* 46 (2025) e70112. https://doi.org/10.1002/jcc.70112.

39. E.S. Kolesnikov, Y. Xiong, A.V. Onufriev, Implicit Solvent with Explicit Ions Generalized Born Model in Molecular Dynamics: Application to DNA, *Journal of Chemical Theory and Computation* 20 (2024) 8724–8739. https://doi.org/10.1021/acs.jctc.4c00833.
40. H.-X. Zhou, X. Pang, Electrostatic Interactions in Protein Structure, Folding, Binding, and Condensation, *Chemical Reviews* 118 (2018) 1691–1741. https://doi.org/10.1021/acs.chemrev.7b00305.
41. D. Tolasa, Molecular Dynamics Simulations: Unraveling the Complexities of Chemical Reactions at the Atomic Level, *American Journal of Physics and Applications* 13 (2025) 46–58. https://doi.org/10.11648/j.ajpa.20251303.11.
42. N. Thomas, T. Sanyal, P. Greisen, K. Deibler, Structure-Based Computational Scanning of Chemical Modification Sites in Biologics, *ACS Omega* 9(34) (2024) 36787–36794. https://doi.org/10.1021/acsomega.4c05857.
43. P. Sangavi, S. Muthu, J.N. Cheerlin Mishma, K. Langeswaran, Vibrational, Theoretical Characterisation and *In Silico* Investigation of Norcinnamolaurine: A Natural Lead for Inhibiting AURKA and Overcoming Immune Evasion in Triple-Negative Breast Cancer, *Molecular Simulation* 50 (2024) 1614–1630. https://doi.org/10.1080/08927022.2024.2420724.
44. T. Kameda, A. Awazu, Y. Togashi, Molecular Dynamics Analysis of Biomolecular Systems Including Nucleic Acids, *Biophysics and Physicobiology* 19 (2022) e190027. https://doi.org/10.2142/biophysico.bppb-v19.0027.
45. C. Guillén-Pingarrón, P.M. Guillem-Gloria, A. Soni, G. Ruiz-Gómez, M. Augsburg, F. Buchholz, M. Anselmi, M.T. Pisabarro, Conformational Dynamics Promotes Disordered Regions from Function-Dispensable to Essential in Evolved Site-Specific DNA Recombinases, *Computational and Structural Biotechnology Journal* 20 (2022) 989–1001. https://doi.org/10.1016/j.csbj.2022.01.010.
46. M. Kožić, B. Bertoša, Trajectory Maps: Molecular Dynamics Visualization and Analysis, *NAR Genomics and Bioinformatics* 6 (2024) lqad114. https://doi.org/10.1093/nargab/lqad114.
47. A. Prajapati, R. Kachhadiya, A. Pravin, S.K. Singh, H. Patel, Computational Insights into Marine-Derived pVHL-HIF1 α Inhibitors: Pharmacophore Modelling, Molecular Docking, MD Simulations, MM/PBSA and DFT Studies, *Molecular Simulation* 51 (2025) 1036–1053. https://doi.org/10.1080/08927022.2025.2574641.
48. Y. Maruyama, R. Igarashi, Y. Ushiku, A. Mitsutake, Analysis of Protein Folding Simulation with Moving Root Mean Square Deviation, *Journal of Chemical Information and Modeling* 63 (2023) 1529–1541. https://doi.org/10.1021/acs.jcim.2c01444.
49. C.C. David, D.J. Jacobs, Principal Component Analysis: A Method for Determining the Essential Dynamics of Proteins, in: Livesay, D. (eds), *Protein Dynamics. Methods in Molecular Biology*. Humana Press, Totowa, NJ, 2014: pp. 193–226. https://doi.org/10.1007/978-1-62703-658-0_11.
50. G.G. Maisuradze, A. Liwo, H.A. Scheraga, Principal Component Analysis for Protein Folding Dynamics, *Journal of Molecular Biology* 385 (2009) 312–329. https://doi.org/10.1016/j.jmb.2008.10.018.

51. M.A. Pravin, S.K. Singh, Enhanced Sampling and Free Energy Methods to Study Protein Folding and Dynamics, in: Saudagar, P., Tripathi, T. (eds), *Protein Folding Dynamics and Stability*. Springer, Singapore, 2023: pp. 165–179. https://doi.org/10.1007/978-981-99-2079-2_9.
52. R.C. Bernardi, M.C.R. Melo, K. Schulten, Enhanced Sampling Techniques in Molecular Dynamics Simulations of Biological Systems, *Biochimica et Biophysica Acta (BBA)—General Subjects* 1850 (2015) 872–877. https://doi.org/10.1016/j.bbagen.2014.10.019.
53. J. Hénin, T. Lelièvre, M.R. Shirts, O. Valsson, L. Delemotte, Enhanced Sampling Methods for Molecular Dynamics Simulations [Article v1.0], *Living Journal of Computational Molecular Science* 4 (2022) 1583. https://doi.org/10.33011/livecoms.4.1.1583.
54. A. Ghidini, E. Serra, A. Cavalli, On Free Energy Calculations in Drug Discovery, *Accounts of Chemical Research* 58 (2025) 3137–3145. https://doi.org/10.1021/acs.accounts.5c00465.
55. M. Arun Pravin, C. Selvaraj, S.K. Singh, A Multiscale *In Silico* Approach to Impede the Interaction Between Dengue NS1 and Human RPL18, Aiming to Suppress the Dengue Viral Translation and Proliferation, *Molecular Simulation* 50 (2024) 1642–1665. https://doi.org/10.1080/08927022.2024.2426587.
56. E. King, E. Aitchison, H. Li, R. Luo, Recent Developments in Free Energy Calculations for Drug Discovery, *Frontiers in Molecular Biosciences* 8 (2021) 712085. https://doi.org/10.3389/fmolb.2021.712085.
57. M.R. Challapa-Mamani, E. Tomás-Alvarado, A. Espinoza-Baigorria, D.A. León-Figueroa, R. Sah, A.J. Rodriguez-Morales, J.J. Barboza, Molecular Docking and Molecular Dynamics Simulations in Related to Leishmania donovani: An Update and Literature Review, *Tropical Medicine and Infectious Disease* 8 (2023) 457. https://doi.org/10.3390/tropicalmed8100457.
58. M. Ahmed, A.M. Maldonado, J.D. Durrant, From Byte to Bench to Bedside: Molecular Dynamics Simulations and Drug Discovery, *BMC Biology* 21 (2023) 299. https://doi.org/10.1186/s12915-023-01791-z.
59. K. Monisha, S. Mahema, M. Chokkalingam, S.F. Ahmad, T.B. Emran, P. Prabu, S.S. Ahmed, Elucidating the Histone Deacetylase Gene Expression Signatures in Peripheral Blood Mononuclear Cells That Correlate Essential Cardiac Function and Aid in Classifying Coronary Artery Disease Through a Logistic Regression Model, *Biomedicines* 11 (2023) 2952. https://doi.org/10.3390/biomedicines11112952.
60. J. Sudhan, S.F. Ahmad, A. Wani, S.S. Ahmed, Phytochemicals from *Piper betle* (L.) as Putative Modulators of a Novel Network-Derived Drug Target for Coronary Artery Disease: An *In Silico* Study, *Processes* 11 (2023) 3064. https://doi.org/10.3390/pr11113064.
61. V. Janakiraman, M. Sudhan, F.A. Sheikh, S.S. Shiek, Pharmacoscreening, Molecular Dynamics, and Quantum Mechanics of Inermin from Panax Ginseng: A Crucial Molecule Inhibiting Exosomal Protein Target Associated with Coronary Artery Disease Progression, *PeerJ* 11 (2023) e16481. https://doi.org/10.7717/peerj.16481.

62. V. Janakiraman, M. Sudhan, S.F. Ahmad, S.M. Attia, T.B. Emran, S.S. Ahmed, Molecular Docking, Quantum Mechanics and Molecular Dynamics Simulation of Anti-CAD Drugs Against High-Risk Xanthine Dehydrogenase Variants Associated with Oxidative Stress Pathways, *Journal of Computational Biophysics and Chemistry* 23 (2024) 1109–1128. https://doi.org/10.1142/S2737416524500315.
63. M. Sudhan, V. Janakiraman, S.F. Ahmad, S.M. Attia, R. Subramanian, D. Devi, S.S. Ahmed, A Comprehensive Insight from Molecular Docking and Dynamics with Clinical Investigation on the Impact of Direct Oral Anticoagulants on Atheroprotective Protein in Atrial Fibrillation, *BMC Pharmacology and Toxicology* 25 (2024) 56. https://doi.org/10.1186/s40360-024-00785-z.
64. A. Ocana, A. Pandiella, C. Privat, I. Bravo, M. Luengo-Oroz, E. Amir, B. Gyorffy, Integrating Artificial Intelligence in Drug Discovery and Early Drug Development: A Transformative Approach, *Biomarker Research* 13 (2025) 45. https://doi.org/10.1186/s40364-025-00758-2.
65. V. Patel, M. Shah, Artificial Intelligence and Machine Learning in Drug Discovery and Development, *Intelligent Medicine* 2 (2022) 134–140. https://doi.org/10.1016/j.imed.2021.10.001.
66. A. Pathak, R. Theagarajan, M.M. Rizqi, A.S. Nugraha, T. Boruah, Pratiksha, H. Kumar, B. Naik, S. Yadav, A.K. Jha, A. Trivedi, A.K. Gupta, P. Kumar, AI-Enabled Drug and Molecular Discovery: Computational Methods, Platforms, and Translational Horizons, *Discover Molecules* 2 (2025) 32. https://doi.org/10.1007/s44345-025-00037-5.

Chapter 2

In silico ADMT prediction

Mutthuraj Dasegowda[1], H. D. Sahana[2], Mouvanal Sajana[1], Moganamurthy Brunda[1], Prasad Tejas[3], and Kanthesh M. Basalingappa[1]

[1]Division of Molecular Biology, SLS, JSS Academy of Higher Education & Research, Mysuru, Karnataka, India

[2]Department of Biochemistry, Oxford Medical College and Hospital, Bengaluru, Karnataka, India

[3]Department of Biochemistry, JSS Medical College, JSS Academy of Higher Education & Research, Mysuru, Karnataka, India

2.1 INTRODUCTION TO IN SILICO ADMET PREDICTION

Pharmacokinetic and safety profile assessment of a compound is a significant factor that defines the success of the compound in a complex and resource-intensive process involved in drug discovery and development. The most important pharmacological characteristics influencing the behavior of a drug in the human organism are encoded in the acronym ADMET, which stands for Absorption, Distribution, Metabolism, Excretion, and Toxicity. The ADMET profiling has traditionally been based on experimental techniques that are labor-intensive and expensive, often involving in vivo animal tests and in vitro assays. The development of computational (in silico) methods has changed all of this, and scientists can now forecast the ADMET properties early in the drug development process in an unprecedented efficient and precise manner [1]. In silico ADMET prediction is based on using computational systems to predict the absorption of a compound into the bloodstream, its distribution in tissues, its metabolism by enzymes, its excretion by the body as well as its toxicity potential. These predictive technologies have become invaluable to contemporary pharmacology, as they limit the use of animal experimentation, save on expenses, and speed up the process of determining valuable drug candidates. In this chapter, the author gives a general overview of in silico ADMET prediction, its definition, scope, importance, historical development, and the wide variety of computational tools that exist, including early rule-based systems as well as more recent artificial intelligence (AI)-based models [2]. The role of the approaches as being applicable to the contemporary issues of drug

DOI: 10.1201/9781003672760-2

discovery and the contribution to enhancing the success rates of clinical development are also highlighted in the chapter.

2.2 DEFINITION AND SCOPE OF ADMET

ADMET consists of five important pharmacokinetic and pharmacodynamic properties that define the efficacy and safety of a drug:

1. **Absorption**: This means how a drug is absorbed into the bloodstream by its site of administration (e.g., oral, intravenous or transdermal route). The determination of solubility, permeability through biological membranes, and transporter existence of P-glycoprotein, among others, are key determinants of absorption.
2. **Distribution**: A drug once absorbed is distributed to different tissues and organs. Factors affecting distribution include plasma–protein binding, tissue affinity, and blood-brain barrier (BBB) penetration that affect the bioavailability of the drug and its therapeutic penetration.
3. **Metabolism**: This is the enzymatic conversion of a drug to metabolites, mainly in the liver, by means of oxidation, reduction, or hydrolysis. Cytochrome P450 enzymes are very important in the metabolism, which may activate or inactivate a drug or generate some toxic products.
4. **Excretion**: The excretion of a drug or its metabolites by the body, usually through the kidneys, though also by bile, lungs, or sweat, is the excretion profile of a drug. Important parameters in this process are clearance rates and half-life.
5. **Toxicity**: This is the negative impact that a drug can have on biological systems, which include mild side effects to extreme organ toxicity or carcinogenicity. Some of the toxicity assessments involve acute, chronic, and genotoxic effects and drug–drug interactions [3].

ADMET is a very wide field because it incorporates multidisciplinary expertise that incorporates chemistry, biology, pharmacology, and toxicology. It is very important that ADMET profiling be done properly because a drug candidate should not only be effective; it must also be safe to use by humans. In silico ADMET prediction builds this by offering the computational facilities to make such predictions so that the early prediction of potential problems can be identified before expensive experimental investigations are pursued [3,4].

2.3 IMPORTANCE IN DRUG DISCOVERY AND DEVELOPMENT

The low success rates in drug development, with only a small percentage of drug development candidates making it through preclinical development to successful approval in the market, highlight the significance of early and precise ADMET evaluation. In the past, poor ADMET profiles, especially poor pharmacokinetics (PK) or unexpected toxicity, have led to up to 40%–60% drug failures in clinical trials. ADMET prediction in in silico is able to counter these hurdles by providing a number of advantages, which include the following:

1. **Cutting the Attrition Rates**: Predicting the ADMET properties at either the hit-to-lead or lead optimization steps, in silico tools can be used to filter out compounds with poor profile, and hence more likely to succeed in the clinical phase. As an example, inefficient oral bioavailability or excessive toxicity should be predicted early enough to make structural changes to favor a compound.
2. **Cost Efficiency**: It is a costly venture to undertake the drug development process, and the cost of approved drugs is usually over a billion dollars. Prediction of ADMET in silico saves time and resources because one does not need to conduct a lot of in vitro or in vivo testing. Computational models can screen thousands of compounds virtually, and those with the most desirable ADMET properties can then be developed [5].
3. **Ethical Implications**: Traditional ADMET profiling is still based on animal testing and this raises the ethical issue of animal welfare. In silico techniques fit the concept of the 3Rs (Replacement, Reduction, and Refinement) in that minimal animal testing is done by virtual screening and prediction of drug activity, thereby promoting more humane drug development strategies.
4. **Rapid Development Scheduling**: By incorporating ADMET predictions into the drug discovery process at the initial stage, scientists will be able to make informed choices in a shorter period of time, reducing the overall development schedule. This is especially important in urgent medical situations, such as during medical pandemics or for rare diseases.

The in silico prediction of ADMET is considered the pillar of contemporary drug discovery, as it allows implementing a proactive as opposed to reactive approach to the resolution of pharmacokinetic and safety issues [6].

2.4 EVOLUTION FROM TRADITIONAL EXPERIMENTAL METHODS TO COMPUTATIONAL APPROACHES

The development of the ADMET profiling is indicative of the general improvement of science and technology. Conventional experimental technologies, created in the middle of the 20th century, were based on the empirical evidence of in vitro tests (e.g., cellular permeability of Caco-2 cells in the context of absorption) and in vivo tests in animals (e.g., rodents in the context of toxicity). Although these techniques were useful, they were restricted due to high cost, time consumption, interspecies differences, and ethical issues. With the advent of computational chemistry in the 1980s, the situation changed. The initial in silico techniques were quantitative structure-activity relationship (QSAR) models, which were statistical regressions of biological activity versus molecular structure. These models employed variables of molecular weight (MW), hydrophobicity (logP) and polar surface area to determine ADMET properties. As an example, the simple heuristic of Lipinski, the Rule of Five, which was first proposed in 1997, gave a rule-of-thumb method of oral bioavailability based on molecular properties and gave the basis to rule-based ADMET prediction [7].

In the 2000s, more sophisticated tools were created due to the progress in computational power and cheminformatics. Detailed predictions of drug-enzyme interactions and membrane permeability were made using molecular dynamics (MD) simulations, docking studies, and pharmacophore modeling. The further development of machine learning (ML) in the 2010s changed the situation in a different way, as it became possible to analyze vast quantities of data to draw intricate patterns in ADMET behavior. Nowadays, models based on AI, such as deep learning and neural networks, provide accuracy never seen before in predicting a particular property, such as cytochrome P450 metabolism or cardiotoxicity [8].

This development has transformed ADMET profiling into a predictive method rather than an experimental reactive process, and in silico tools have become part and parcel of all drug discovery processes.

2.5 OVERVIEW OF IN SILICO TOOLS: FROM RULE-BASED SYSTEMS TO AI-DRIVEN MODELS

In silico prediction of ADMET includes a range of computational tools, each with different methodologies and applications:

1. **Rule-Based Systems**: These are based on empirical rules that are based on experimental findings. As an example, the Lipinski Rule of Five is used to predict oral bioavailability depending on the MW (less than

500 Da), logP (less than 5), the number of hydrogen bond donors (less than 5), and the number of hydrogen acceptors (less than 10). Equally, the rules proposed by Veber determine bioavailability based on polar surface area and rotatable bonds. Just as simple and interpretable, rule-based systems are constrained by having predefined thresholds and being unable to determine complex biological interactions.
2. **QSAR Models**: QSAR models are statistical models that correlate molecular descriptors with the ADMET properties. Recent models of QSAR use thousands of descriptors, such as topological, electronic, and steric ones, to predict such outcomes as solubility or toxicity. These models are very popular because they are accurate and efficient in computation.
3. **Molecular Modeling and Simulations**: Some of the techniques used in the simulation of drug interaction are molecular docking and dynamics simulations, which model drug interactions at the atomic level. As an example, the docking studies can forecast the affinity to the cytochrome P450 enzymes, and the MD can be used to simulate the drug transport across the membranes. Such approaches are computationally expensive yet give high-resolution information.
4. **ML Models**: ML algorithms, for example, random forests and support vector machines (SVM), utilize big data to forecast ADMET properties. These models are very effective in determining non-linear relationships and find a wide application in activities such as toxicity prediction or modeling drug–drug interactions.
5. **AI-Driven Models**: The newest trends include deep learning and neural networks, which utilize large datasets of available databases (e.g., PubChem, ChEMBL) to model the prediction of ADMET properties with high accuracy. As an example, graph neural networks (GNNs) can be used to simulate the structure of molecules as a graph, which represents complex interactions between chemicals. The AI models are mostly useful in making predictions of complex endpoints, which include idiosyncratic toxicity or BBB penetration [9–12].

2.6 OBJECTIVES OF THE CHAPTER AND ITS RELEVANCE TO MODERN PHARMACOLOGY

The chapter will primarily focus on introducing in silico ADMET prediction to give the reader a clear picture of the concept, processes, and uses of the prediction. They are:

1. Establishing the definition of the importance of ADMET in drug discovery.
2. Emphasizing the in silico practical and ethical advantages.

3. The history of the development of experimental and computational approaches.
4. A review of in silico tools and their strengths.
5. To prove the applicability of in silico ADMET prediction in solving the current issues in the field of pharmacology.

The topicality of the chapter to the contemporary pharmacology is hard to be overestimated. In silico ADMET prediction is a disruptive solution to rising demands in the pharmaceutical industry to produce safe and effective drugs in less time and at a lower cost. These tools provide an opportunity to minimize the rate of attrition and the allocation of resources according to priorities, as well as facilitate ethical development of drugs by ensuring that the ADMET liabilities are identified at an early stage. Moreover, the application of AI and AI-based big data analytics to ADMET prediction is in line with the wider movement toward precision medicine and personalized therapies that require the highest possible accuracy of pharmacokinetic and safety profiles [13]. To sum up, in silico ADMET prediction is an innovation in drug discovery, which is poised to bridge chemistry, biology, and data science to speed up the discovery of life-saving therapies. This chapter offers a baseline to researchers and practitioners that would like to make use of these tools to promote pharmacological innovation.

2.7 THEORETICAL FOUNDATIONS OF IN SILICO ADMET PREDICTION

In silico prediction of ADMET has transformed the drug discovery process in that it allows the prediction of the pharmacokinetic and safety profile of a compound with a computational model at a very early stage of the drug discovery process. This strategy is based on a sound theoretical framework that incorporates computational chemistry, PK, systems biology, and regulatory science. In silico tools decrease expenses by modeling the intricate biological mechanisms and have lowered the amount of animal testing and shortened the length of time it takes to find a viable drug candidate. This part discusses the theoretical background of in silico ADMET prediction, underlying principles of computational chemistry and PK, use of molecular descriptors, Quantitative Structure-Property Relationships (QSPR) and QSAR, interaction with systems biology and PK, and an analysis of regulatory guidelines provided by Food and Drug Administration (FDA), European medicines agency (EMA), and ICH. It has four tables that summarize the key methods, descriptors, models, and guidelines that facilitate the ease with which this information can be accessed by researchers and practitioners in the field of modern pharmacology [14].

2.7.1 Basic principles of computational chemistry and pharmacokinetics

In silico ADMET prediction is scientifically justified by the use of computational chemistry and PK, which makes it possible to model the interactions of molecules and the disposition of drugs occurring in the biological environment.

2.7.1.1 Computational chemistry

Computational chemistry is the field that applies mathematical algorithms and computational methods to the study of chemical systems to provide a better understanding of how to determine the properties of molecular systems in ADMET. Key methods include the following:

1. **Quantum Mechanics (QM)**: QM techniques, including Hartree-Fock, density functional theory (DFT), or semi-empirical techniques, are used to compute electronic structures used to predict properties such as ionization potential, electron affinity, or chemical reactivity. These are essential for modeling metabolism (e.g., cytochrome P450 interactions) or toxicity (e.g., reactive metabolite formation). The computational cost of QM is too high such that it can only be used with small molecules or in high-precision applications.
2. **Molecular Mechanics (MM)**: MM treats molecules as assemblies of atoms where study is done in a classical mechanics framework, with bond lengths, angles, and non-bonded interactions (e.g., van der Waals forces) being represented using force fields (e.g., AMBER, CHARMM). MM has found extensive application in MD simulations of drug–membrane interactions (absorption) or of protein–ligand binding (distribution).
3. **Cheminformatics**: It is a field of study that involves the analysis of chemical data with using tools like RDKit, ChemAxon, or OpenBabel. Cheminformatics enables the screening of compound libraries on a large scale and predicting ADMET properties based on structural features such as molecular mass or lipophilicity (Table 2.1).
4. **Hybrid QM/MM Methods**: Hybrid approaches combine the precision of QM with the efficiency of MM. These approaches can be used to study in detail the interaction between enzymes and their substrates during metabolism or toxicity.
5. **Molecular Docking**: This method predicts the binding orientation and affinity of a drug to a target protein, which is useful in metabolism (e.g., cytochrome P450 binding) or toxicity (e.g., hERG channel inhibition) [15,16].

Table 2.1 Key computational chemistry methods for ADMET prediction [16]

Method	*Description*	*ADMET application*	*Advantages*	*Limitations*
Quantum mechanics	Computes electronic structures based on quantum theory (e.g., DFT, Hartree-Fock)	Predicts metabolism, toxicity, reactivity	Extremely high precision for small systems	Computer intensive, confined to small molecules
Molecular mechanics	Models interactions in a classical field using force fields and classical physics	Mimics drug–membrane or protein–ligand binding	Rapid, can be used in large systems	Less accurate for electronic properties
Cheminformatics	Calculates descriptors and performs virtual screening on chemical data	Predicts ADMET characteristics through structural characteristics	User-friendly, high-throughput tools	Constrained by the quality of descriptors
Hybrid QM/MM	Integrates QM and MM to obtain a multi-scale model	Enzyme–substrate interrelations	Accurate and efficient	Complex installation, requires expertise
Molecular docking	Binds affinity and prediction of orientation	Determines metabolism, toxicity, or location	Fast, widely applicable	Limited by scan ability

2.7.1.2 Pharmacokinetics

PK characterizes the time-varied drug absorption, diffusion, metabolism, and discharge in the body. These processes can be modeled in silico as ADMET prediction:

1. **Absorption**: Models simulate the rate and degree of drug uptake, relying on parameters such as membrane permeability (depending on lipophilicity or polar area) and transporter activity (e.g., P-glycoprotein, organic anion transporters). For example, computational tools can be used to estimate intestinal absorption using LogP and solubility.
2. **Distribution**: The In silico models predict tissue partitioning, plasma–protein binding, and penetration through cellular barriers such as the BBB. Molecular docking and MD simulations have been used to

predict binding to albumin or BBB permeability, which affects the volume of distribution (Vd).

3. **Metabolism**: Simulation of enzymatic transformations, especially of cytochrome P450 enzymes, which metabolize about three-quarters of drugs, is done by computational tools. Reactive sites are determined using quantum mechanical calculations, whereas metabolic pathways and toxic metabolites are predicted using ML.
4. **Excretion**: Using physicochemical properties (e.g., solubility, MW) and physiological parameters (e.g., glomerular filtration rate), models are used to estimate the clearance rates and half-life. Physiologically based pharmacokinetic (PBPK) modeling is a combination of these variables used to determine renal or biliary excretion [17].

The principles allow the in silico tools to model the complex pharmacokinetic processes, which offer a mechanistic foundation for predicting ADMET.

2.7.2 Molecular descriptors and their role

Molecular descriptors are numerical numbers that describe the physicochemical, structural, topological, or electronic features of a molecule. They can be used as inputs in the predictive ADMET models, which include aspects that could affect both the pharmacokinetic and toxicological effects. Key descriptors include the following:

2.7.2.1 Physicochemical descriptors

1. **LogP (Partition Coefficient)**: An indicator of lipophilicity, it represents the solubility of octanol and water. An increase in the values of LogP is an increase in absorption but toxicity (e.g., hepatotoxicity) can occur.
2. **Topological Polar Surface Area (TPSA)**: Measures polar surface area, which influences the hydrogen bonding and crossing the membrane barrier. TPSA below 140 A 2 is predictive of good oral bioavailability, whereas TPSA below 60 A 2 is predictive of BBB penetration.

2.7.2.2 Structural descriptors

Hydrogen Bond Donors/Acceptors: LETE solubility and permeability. The number of donors (>5) or acceptors (>10) can be too large to allow absorption.

Rotatable Bonds: These are evidence of molecular flexibility which influences bioavailability and binding to receptors. They also prefer fewer rotatable bonds (less than 10) when it comes to oral drugs.

2.7.2.3 Topological descriptors

1. **Wiener Index**: Molar shape and connectivity—Capture and change solubility or binding with protein.
2. **Shape Descriptors**: Characterize 3D geometry, which is pertinent with regard to the receptor interactions.

2.7.2.4 Electronic descriptors

1. **Partial Charges, Dipole Moments**: Govern interaction with an enzyme or a receptor and this affects metabolism and toxicity.
2. **HOMO-LUMO Gap (Chemical Reactivity)**: It represents how reactive the molecule is, which can be used to predict metabolic stability, domestication, and electrophilicity.

This is done by obtaining descriptors based on software such as RDKit, ChemAxon, Dragon, or PaDEL, and thousands of features in each molecule are produced. Descriptors are associated with experimental data to provide model development in ADMET prediction. For example, a model involving LogP, TPSA, and MW can be used to predict oral bioavailability, and electronic descriptors are used to detect reactive sites that are likely to be toxic. The selection (Table 2.2) of features (e.g., recursive feature elimination) may ensure that only relevant features are described to improve the robustness of the model [19].

Table 2.2 Common molecular descriptors in ADMET prediction [19]

Descriptor type	*Examples*	*ADMET relevance*	*Typical range for drug-likeness*
Physicochemical	LogP, TPSA, Molecular Weight	Absorption, distribution, solubility	LogP: 0–5, TPSA: <140 Å², MW: <500 Da
Structural	H-bond donors/ acceptors, Rotatable bonds	Absorption, bioavailability	Donors: <5, Acceptors: <10, Rotatable: <10
Topological	Wiener Index, Balaban Index	Solubility, protein binding	Context-dependent
Electronic	Partial charges, HOMO-LUMO gap	Metabolism, toxicity	Context-dependent

2.7.3 Quantitative structure-property relationships (QSPR) and quantitative structure-activity relationships (QSAR)

QSAR and QSPR are statistical modeling schemes to correlate molecular descriptors with physicochemical characteristics (QSAR) or biological activities (QSPR), especially to support the predictable in silico prediction of ADMET.

2.7.3.1 QSPR

QSPR models offer predictive capabilities of physicochemical properties that are important to ADMET, including solubility, lipophilicity, or ionization state. To give an example, a QSPR model could incorporate MW, LogP, and TPSA in not only predicting aqueous solubility, which is important in oral absorption, but also in predicting other determinants of oral absorption, such as solubility in other solvents. Common methods include:

1. **Multiple Linear Regression (MLR)**: Predicts relationships that form linear attempts with descriptors and properties, which will be utilized to create interpretable non-opaque models.
2. **Partial Least Squares (PLS)**: It is able to work with high-dimensional sets of descriptors and mitigates overfitting with complex data.
3. **ML**: Given the random forest or gradient boosting algorithms or any other neural network, they enhance non-linear relations [20].

2.7.3.2 QSAR

QSAR models are used to make predictions associated with biological activities, for example, enzyme inhibition, receptor binding, or toxicity. These can include hERG channel inhibition (associated with cardiotoxicity), which is ominously predicted by a QSAR model with electronic descriptors and steric descriptors. Contemporary QSAR methods include the following:

1. **Classical QSAR**: MLR or PLS are interpretable predictors, and they are applicable to small data.
2. **ML QSAR**: This involves using algorithms such as SVM, random forests, or deep neural networks to learn more complicated patterns in large data sets.
3. **Deep Learning QSAR**: GNNs (model molecules as graphs) are good at making enhanced predictions of complex endpoints such as idiosyncratic toxicity or penetration of a drug into the BBB [20,21].

Table 2.3 QSPR and QSAR model types for ADMET prediction [22]

Model type	*Methodology*	*ADMET application*	*Advantages*	*Limitations*
Classical QSPR/QSAR	MLR, PLS	Solubility, lipophilicity, enzyme inhibition	Interpretable, computationally efficient	Limited to linear relationships
Machine learning QSAR	Random forests, SVM, gradient boosting	Toxicity, bioavailability, drug interactions	Handles non-linear relationships	Requires large datasets, less interpretable
Deep learning QSAR	Neural networks, GNNs	Complex endpoints (e.g., idiosyncratic toxicity)	High accuracy for complex data	Computationally intensive, black-box models

2.7.3.3 *Applications/major problems*

QSPR/QSAR models are used to drive such high-throughput ADMET property screening tools as ADMETlab, SwissADME, and pkCSM. Its uses are the prediction of cytochrome P450 inhibition, genotoxicity, or oral bioavailability (Table 2.3). Challenges include the following:

1. **Data Quality**: The models need quality and variation of training data. Limited generalizability is a result of sparse or biased ODL datasets (e.g., low chemical diversity).
2. **Oversizing**: This occurs when an excessive number of descriptors are used and end up, leading to overfitting and resulting in an erroneous predictor.
3. **Applicability Domain**: Maturated compounds in a space not included in the chemical space of their training set may receive faulty predictions, requiring sound validation (i.e., cross-validation, external test sets).
4. **Interpretability**: State-of-the-art ML models (e.g., deep learning) can be not transparent, which makes it difficult to accept in regulations.

Recent developments, including ensemble modeling, transfer learning, and combination of multi-omics data, have improved accuracy and fine body of QSPR/QSAR [22].

2.7.4 Integration with systems biology and pharmacodynamics

ADMET prediction is increasingly being combined with systems biology and PK to predict drug behavior in complex biological systems more accurately and improve predictivity and therapeutic utility.

2.7.4.1 IB systems bio thesis

Biological network models integrate interactions across metabolic pathways, signaling cascades, and organ systems to provide a holistic understanding of ADMET processes:

PBPK Modeling: PBPK models are used to simulate organ compartments (e.g., blood, liver, and kidneys) of drug disposition with physiological parameters (e.g., organ volumes, blood flow) and ADMET properties. To provide an example, PBPK models can be utilized to generate predictions of drug concentrations in tissues of interest, which is useful for optimizing doses and evaluating drug–drug interactions. Tools such as Simcyp or GastroPlus combine ADMET and PBPK data [23].

Metabolic Network Analysis: Computers map out metabolic pathways, including essential enzymes (e.g., CYP3A4, UGTs) and any possible associations. Network-based model predicts the influence of inhibition of one of the enzymes in the metabolism of other drugs given concomitantly (essential in the case of polypharmacy).

Multi-scale Modeling: Merges between molecular-scale data (e.g., QSAR predictions) and organ- and organism-scale data (e.g., PBPK models) with scaling data between the atomic and molecular level (i.e., lung saturation) on the one hand (ADMET) and macro-level biological equilibrium (i.e. PK) on the other hand (i.e., lung saturation). An example is the use of multi-scale models to forecast the influence of physicochemical characteristics of a drug on the systemic clearance or tissue-specific toxicity [24].

2.7.4.2 Pharmacodynamics integration

PD is a term used to describe the effects that a drug has on the body such as the effectiveness of the drug and toxicity. Combination of ADMET and PD models allows for improved prediction. PK is connected to biological response:

1. **Efficacy Prediction**: Molecular docking and QSAR models are predictive of receptor binding affinity, which matches drug distribution parameters with therapeutic effects. The docking assays give an example and estimate a drug to have potency against a target kinase or GPCR.
2. **Toxicity Prediction**: PD models are used to predict toxic events, which could include hERG channel and/or hepatotoxicity, using a combination of ADMET data and molecular interaction. ADMET and PD are incorporated in ML models to offer dose-sensitive toxicity.
3. **PK/PD Modeling**: Integrates pharmacodynamic (e.g., dose–response curves) and pharmacokinetic (e.g., drug concentration-time profiles) information to determine a dose regimen that is

optimal. As an example, PK/PD is useful in predicting the safety margin of drugs with slim therapeutic indexes, as is the case with chemotherapeutics [25].

This combination allows one to achieve a systems-level perception of drug behavior, which can be used in the process of designing compounds with matched efficacy and safety profile.

2.7.5 Regulatory perspectives: guidelines from FDA, EMA, and ICH

In silico prediction of ADMET in drug development through in silico methods is accepted by regulatory organs, including the FDA, EMA, and International Council for Harmonization of Technical Requirements for Pharmaceuticals for Human Use (ICH). Their guidelines (Table 2.4) offer guidelines on how computational models should be incorporated in submissions under regulations with a key focus on validation, transparency, and applicability [26].

FDA: The Model-Informed Drug Development (MIDD) program was initiated in the year 2018 at FDA and encourages the use of in silico tools (such as QSAR, QSPR, and PBPK models) to facilitate drug development and regulatory decision-making. It is recommended that physiologically based pharmacokinetic modeling be developed to employ the model to predict drug interactions, dosage, and special population PK (e.g., in pediatrics or in renally impaired patients). The FDA guidelines in 2018, PBPK

Table 2.4 Regulatory guidelines for in silico ADMET prediction [30]

Agency	*Guideline*	*Key focus*	*Requirements*
FDA	MIDD Program (2018), PBPK Analyses Guidance (2018)	PBPK, QSAR for PK, toxicity, drug interactions	Model validation, sensitivity analysis, transparent documentation
EMA	PK/PD Modeling Guideline (2016), PBPK Reporting Guideline (2018)	QSAR, PBPK for trial design, safety assessment	Validation against experimental data, applicability domain assessment
ICH	M7 (2014, rev. 2020), S9 (2009), Q3D (2014, rev. 2020)	QSAR for genotoxicity, nonclinical toxicity in oncology, elemental impurities	Two QSAR methods for genotoxicity, expert review, integration with experimental data
General	Cross-agency emphasis on 3Rs, transparency	Ethical drug development, model reliability	Clear reporting of model assumptions, limitations, and validation metrics

Analyses—Format and Content, specify how such models can be formulated, validated, and applied for dose predictions and the PK of special patients. As another part of its 2020 guidance on toxicity prediction of impurities, the FDA endorses the application of QSAR, especially in predicting toxicity in drug or biologic combinations [27].

EMA: EMA takes the position in favor of in silico modeling in optimizing trial design and evaluating safety by their guideline on the Use of Pharmacokinetic and Pharmacodynamic Modeling in Drug Development (2016). The EMA assists with QSAR and PBPK modeling of predicting ADMET, specifically with regard to genotoxicity and drug–drug interactions. The 2018 expression of forms of the EMA includes a guideline on reporting of PBPK modeling and simulation where a model is documented with complete sensitivity analysis, comparison of the model with experimental data, and evaluation of applicability domain [28].

ICH: ICH M7 guideline (2014, revised 2020), Assessment and Control of DNA Reactive (Mutagenic) Impurities in Pharmaceuticals expressly recognizes the use of QSAR models in predicting the genotoxicity of impurities. It involves two complementary QSAR methodologies (e.g., rule-based and statistical) in evaluating mutagenicity with expert review to overcome model limitations. The ICH S9 guideline (2009) advocates the use of in silico techniques in nonclinical toxicity of evaluating oncology drug development in the advancement of computational predictions to minimize animal experimentation. In silico tools can also be used to determine the toxicological risks listed in the ICH Q3D guideline on elements (2014, revised 2020) [29].

The regulatory authorities highlight the following major principles of in silico ADMET models:

1. **Validation**: The model should be tested using measures such as sensitivity, specificity, and concordance with externally tested models to ensure the possibility of generalization.
2. **Transparency**: Model assumptions, training data, and areas of application should be properly recorded so that they can be reviewed by the regulatory authority.
3. **Applicability Domain**: Predictions should be trustworthy for compounds in the chemical space of the model, and the uncertainty on new structures should be determined.
4. **Combination with Experimental Data**: Experimental data must not be used as a substitute rather than an addition to in silico prediction, especially when conditions are seldom associated with cancer, such as carcinogenicity.

Their increasing adoption of in silico tools indicates a potential for better regulatory decision-making and lower costs of research and experimentation, as well as an ethical direction toward the 3Rs (Replacement,

Reduction, Refinement) used in testing on animals. To illustrate the point, the FDA report on MIDD (2021) indicated that PBPK models were applied in more than 250 regulatory submissions during 2018–2020, which illustrates that they affect drug approvals [30].

Theoretical ADMET prediction has theoretical underpinnings consisting of computational chemistry, PK, systems biology, and regulatory science, which are specialists to allow an effective drug discovery framework. Computational chemistry techniques, including QM, MM, and cheminformatics, allow specifically modeling molecular interactions, and the principles of PK are used to convert them into drug disposition prediction. The connection between structure and function is molecular descriptors, as they drive predicting the physicochemical and biological properties using the boxes of QSPR and QSAR. Systems biology and PD integration provide an overall perspective on drug behavior;, regulatory guidelines provided by the FDA, EMA, and ICH guarantee that the in silico predictions contain high standards of reliability and applicability. The additions of tables with the recap of important methods, descriptors, models, and guidelines make the contribution toward the sphere of accessibility. With the ever-changing nature of computational tools and regulatory language, in silico ADMET prediction will be more and more at the forefront when providing safe, effective therapeutics that are efficient and ethical [31].

2.8 IN SILICO ADMET PREDICTION COMPUTATIONAL METHODS AND MODELS

In silico prediction In silico ADMET prediction is based on the wide range of the computational procedures and models to estimate the pharmacokinetic and safety profile of a drug candidate. These methods are based on the theoretical principles of computational chemistry and PK (see previous section), and such methods can allow scientists to filter their compound collections during the preliminary stages of the drug discovery process, thereby driving costs down, animal testing to the minimum, and the development into a shorter period. Here one can find out how computational models and methods are applied to in silico ADMET prediction (e.g., rule-based, methods like the Rule of Five, Veber rules, random forests, and support discharge AI) as well as molecular modeling (e.g., employing docking and MD simulations, QM) and ensemble and consensus modeling as a way of obtaining more accurate predictions and a combination of in silico and in vitro data. In silico ADMET prediction (Figure 2.1) using the combination of those approaches seems to offer a potent predictive framework as applied to optimizing drug candidates, meeting regulatory demands, and developing contemporary pharmacology [32].

This illustration (Figure 2.1) represents the step-by-step computational workflow of ADMET prediction applied in drug discovery nowadays. It starts with the chemical structure of a substance, followed by subsequent

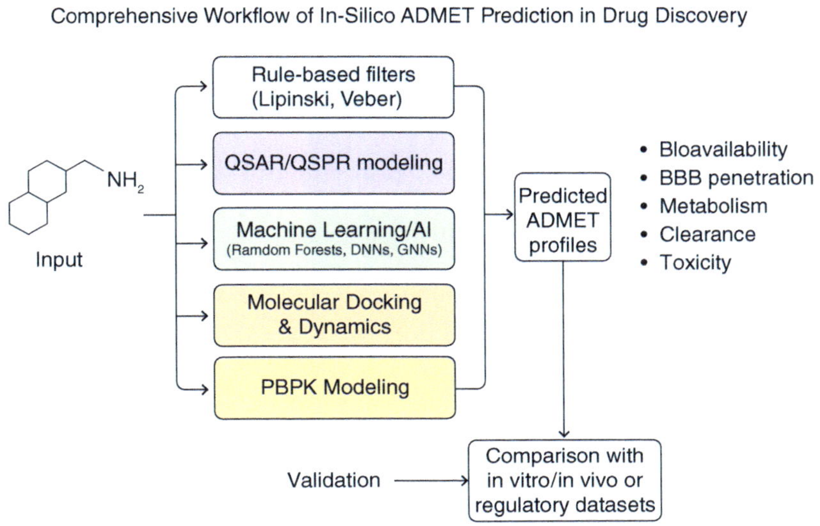

Figure 2.1 Comprehensive workflow of in silico ADMET prediction in drug discovery.

computational tests that assess PK and toxicology characteristics. Key modules include:

1. Filter-based filters (Lipinski rule and Veber rule) of drug-likeness screening.
2. QSAR/QSPR modeling to relate the descriptors of a molecule with its biological activity.
3. ML and AI models (e.g., Random Forests, DNNs, GNNs) of high-accuracy predictions.
4. Mechanistic understanding of mouse protein–ligand interactions by molecular docking and dynamics simulations.
5. PBPK modeling of organ drug distribution and clearance.

The model is tested by comparison of the predicted ADMET profiles (bioavailability, BBB penetration, metabolism, clearance, and toxicity) with in vitro, in vivo, or regulatory data, which also ensures the reliability of the model and translation accuracy in ADMET profiling.

2.8.1 Rule-based approaches

Rule-based methods are easy to interpret, but empirically only methods that predict ADMET properties based on molecular descriptors (Table 2.2 in Section 2.7 provides common molecular descriptors). These algorithms were created based on experimental evidence and are also common in early drug discovery.

2.8.1.1 Lipinski's rule of five

Introduced in 1997, the Lipinski Rule of Five was developed to predict the orally bioavailable form of five physicochemical properties:

1. MW < 500 Da
2. LogP (octanol-water partition coefficient) Less than 5.
3. Hydrogen bond donors < 5
4. Hydrogen bond acceptors < 10

Compounds that do not imply compliance with more than one rule have negative chances of oral absorption. Although it is a simple rule, it works well in a high-throughput screening of compounds, but there is a risk of ruling out viable compounds, like a biologic or active transport of the compound.

2.8.1.2 Veber's rules

The rules by Veber (2002) can be conveniently added to the ones of Lipinski:

1. TPSA less than 140 A2.
2. Rotatable bonds < 10

These regulations focus on membrane permeability and bioavailability and use flexibility and polarity of the molecules. The chemistry compounds that satisfy these criteria ensure high chances of a good oral bioavailability.

2.8.1.3 Other rule-based systems

1. **Ghose Filter**: The rules defined by Lipinski are extended by providing ranges of MW (160–480 Da), LogP (−0.4 to 5.6), molar refractivity (40–130), and the number of atoms (20–70).
2. **Rules in RWCI**: Lead-likeness criteria made more stricter (e.g., MW < 460, LogP < 4.5) to come up with compounds to optimize.
3. **Pfizer's 3/75 Rule**: Infers toxicity potential; the product receives warnings of toxicity based on a LogP over 3, which combined a TPSA lower than 75 A2.

Pros and Cons: Rule-based methods are computationally efficient and understandable and therefore are best used in preliminary screening of the compounds. Nevertheless, they are based on hard cutoffs, can simplify complicated biological processes, and have toned capacity to make predictions over non-oral drugs or non-oral complex outcomes such as toxicity [33–35].

2.8.2 Techniques of machine learning and artificial intelligence

ML and AI methods have revolutionized the in silico ADMET prediction and utilized large datasets in the modeling of non-linear relationships between the characteristics of molecules and ADMET properties (see Table 2.3 in Section 2.7 on the types of QSPR/QSAR models). Such techniques are best at working with high-dimensional data and forecasting complex endpoints such as toxicity or drug interactions [36].

2.8.2.1 Random forests

Random forests are a type of ensemble learning and joins a range of decision trees during prediction of ADMET properties. A subset of descriptors and data are applied to each tree making them less prone to overfitting as well as make them more robust. Random forests have high application in the following:

1. **Absorption**: Prediction of intestinal permeability using eating shipping number LogP and LogPTSA.
2. **Toxicity**: Structural and electronic descriptors of identifying hERG channel inhibition or hepatotoxicity.

2.8.2.2 Support vector machines (SVM)

SVMs are decision trees that learn how to categorize compounds based on a classifier that is optimal (i.e., separating classes) by the hyperplane that maximizes current in high-dimensional space. They are effective for the following:

1. **Metabolism**: Can cytochrome P450 inhibition by computational means using only steric and electronic structural features?
2. **Toxicity**: AS classification according to ICH M7 guidelines: non-mutagenic (or mutagenic).

2.8.2.3 Deep neural networks (DNNs)

Convolutional neural networks (CNNs) and GNNs are examples of DNNs which model complex patterns in molecular data. GNNs, also known as molecules presented in the form of a graph (atoms are represented by nodes and bonds by edges), are effective specifically at:

1. **Complex Endpoints:** The prediction of an idiosyncratic toxicity or blood-brain penetration.
2. **Multi-Task Learning:** Concomitantly predicting various ADMET properties (e.g., absorption and metabolism) based on common network layers [18,37–39].

2.8.2.4 Other AI techniques

1. **Gradient Boosting (MB):** They can be XGBoost/LightGBM, which receive constant model refinements to increase predictive accuracy, such as solubility or clearance models.
2. **Transfer Learning:** Uses the existing trained models on new ADMET endpoints using small volumes of data and improves the prediction of rare toxicity events.
3. **Natural Language Processing (NLP):** Uses ADMET related information found in the literature or databases, which supports the training of models.

Benefits and Shortcomings: ML/AI solutions are highly accurate and adaptable especially when a large dose of data needs to be processed. Nevertheless, they require massive amount of training data, might be uninterpretable (particularly DNNs) and require good validation to guarantee the generalizability within the applicability domain [39].

2.8.3 Molecular modeling

Molecular modeling is a simulation of atomic level of molecular interactions that would give mechanistic information regarding the ADMET processes (computational chemistry methods are present at Table 2.1 of the theoretical basis section). Such techniques are computation-based and have dense predictions.

2.8.3.1 Molecular docking

The method of molecular docking is used to predict drug–protein interaction (confidential: e.g. it is not mirrored by in vivo binding affinity). Applications include the following:

1. **Metabolism:** The ability to predict the binding with the cytochrome P450 enzymes to determine metabolic location.
2. **Toxicity:** Toxicity should be assessed by measuring hERG channel binding to predict cardiotoxicity.
3. **Distribution:** Assessment of plasma–protein binding and albumin.

Computer programs such as AutoDock, Glide, or GOLD evaluate a binding pose using scoring functions to rank binding poses, but accuracy is determined by the quality of protein structure and scoring algorithms [40].

2.8.3.2 Simulations using molecular dynamics (MD)

MD simulations: Simulations of time-dependent behavior of molecular systems (mainly using force fields), for example, AMBER and CHARMM. They are used for the following:

1. **Absorption**: Replicate drug absorption across membranes (e.g., lipid bilayers).
2. **Distribution**: Distribution of study drug across the BBB.
3. **Metabolism**: Research on time dynamics of enzyme substrates.

MD simulations offer finer detail on the changes and interactions of conformations but are computationally intensive.

2.8.3.3 Quantum mechanics (QM)

Electronic properties are calculated using QM and methods such as DFT or semi-empirical solutions to predict reactivity. Applications include the following:

1. **Metabolism**: Localization of reacting sites that are likely to undergo enzyme reaction.
2. **Toxicity**: Preventing action of resulting metabolites that pose a potential threat of genotoxicity.

The hybrid QM/MM approaches combine the accuracy of QM at reaction sites with the efficiency of MM for larger systems to elucidate the details of the environments where enzymes and their substrates interact [40].

Pros and Cons: Molecular modeling is much more insightful and accurate in the particular interactions. Nonetheless, it is computationally complex and demands high-anatropy data and can be not practical in the case of high-throughput screening (HTS).

Ensemble and Consensus Modeling: Ensemble and consensus modeling provides better accuracy and involves the combination of computational models to enhance predictive power and stability. These solutions use the inconclusive powers of individual models to interact with their advantages.

2.8.4 Ensemble and consensus modeling

2.8.4.1 Ensemble modeling

The ensemble modeling combines the prediction of several algorithms (e.g., random forests, SVMs, DNNs) to get one more accurate prediction. Techniques include the following:

Bagging: This is prediction with a combination of model-created predictions made on separate sets of data (e.g., random forests).

Boosting: Improves the models with information after the misprints have been filtered out (e.g., XGBoost).

Stacking: It involves prediction by a meta-model employing various algorithms.

As an illustration, an ensemble model may enable the incorporation of predictions of QSAR with random forests and DNNs to forecast toxicity with better sensitivity and specificity [41].

2.8.4.2 Consensus modeling

Consent modeling averages or votes across the predictions of a multi-task set of several models, eliminating errors. As an example, the ICH M7 guidelines suggest the combination of two application models of QSAR (rule-based and statistical), whose consensus predictions must be confirmed by the review of experts in this field. The approaches that converge find specific usefulness in:

Toxicity: Phototoxicity Integration to determine mutagenicity through task-switching prediction.

Absorption: Characterizing the bioavailability prediction by combining rule-based (e.g., Lipinski) models with ML models.

Merits and Demerits: Ensemble and consensus modeling enhance accuracy and strengths, particularly in relation to complex endpoints. They, however, become more complex in computations and might cause excessive model choice to be redundant [29].

2.8.5 Hybrid data methods in silico and in vitro

Hybrid techniques combine the predictions of these amounts with the experimental data in vitro to serve as the medium of increasing accuracy and reliability to bridge the divide between the computational and experimental methods.

2.8.5.1 Integration strategies

Data Refinement Data are deemed useful and meaningful, and training into models is applied, either in silico models or in vitro data (e.g., Caco-2 permeability tests). As an illustration, QSAR models being trained using in vitro metabolism data are given a higher accuracy in predicting cytochrome P450 inhibition.

PBPK Models Using In vitro Data: PBPK models represent drug disposition using in vitro data (e.g., intrinsic clearance rates) so that special populations (e.g., pediatrics) can be modeled more predictively.

Iterative Optimization: Predictions In silico aids in compound selection to undergo in vitro testing, results of which are used to cycle the model. As an example, it is possible to dock predictions of hERG binding to prioritize patch-clamp assays and use the above QSAR models.

Multi-Scale Modeling: Uses molecular models in vivo (e.g., QM/MM) coding, along with an in vitro binding affinity, to predict multiple baseline ADMET properties [42].

2.8.5.2 Applications

1. **Hybrid Models**: Hybrid models integrate in silico permeability predictions with in vitro Caco-2 quality using the data to correct bioavailability schemes.
2. **Toxicity**: The prediction made by QSAR is confirmed by in vitro genotoxicity tests (e.g., Ames test), which will guarantee that ICH M7 regulations are not violated.
3. **Metabolism**: Predictions of in vitro metabolic clearance improve with in silico predictive data on in vitro microsomal stability.

2.8.5.3 Regulatory considerations

The regulatory bodies, including the FDA and EMA, support the hybrid strategies to make sure it can be reliable (see Table 2.4 in Section 2.7). As an example, FDA (2018) PBPK guidance suggests using an in vitro validation of the model, whereas ICH M7 demands the in vitro validation of the in silico predictions of genotoxicity. Hybrid methods embrace the 3Rs (Replacement, Reduction, Refinement) reduction in animal testing and high predictive power thereof.

Strengths and weak points: Hybrid approaches make these models more accurate and more accepted by the regulations, modernizing the models based on experimental data. They, however, need access to data of high quality in in vitro procedures, which might be expensive and is restricted perhaps by the existence of suitable assays. The computer-aided systems and algorithms of in silico ADMET prediction, including simple rule-of-thumb predictive systems, such as the Rule of Five (Lipinski); more complex and sophisticated AI and software systems; molecular and ensemble models; and hybrid in silico/in vitro approaches, are an efficient drug discovery toolkit. Rule-based systems are easy to use and can be interpreted, whereas

ML/AI methods are able to effectively process sophisticated data. Mechanistic information is offered by molecular modeling and predictive accuracy is improved by ensemble and consensus modeling. The hybrid techniques occupy the space in between computational and experimental methods and should fit the regulatory requirements of FDA, EMA, and ICH. Through these approaches, scientists will be able to streamline drug candidates at an early stage, lower a drug development cost and enhance ethical drug discovery. A combination of these strategies aided by the intellectual basis informed above defines in silico ADMET prediction as one of the pillars of contemporary pharmacology [43,44].

2.9 INDIVIDUAL ADMET PARAMETER FORECASTING

ADMET parameter predictions are also a foundation of in silico drug discovery, allowing scientists to characterize the PK and safety of a drug during an advancement phase in the development pipeline. In silico ADMET prediction streamlines drug candidate selection, lowers the rate of attrition, and accords with the anticipations of regulatory interests by taking advantage of computational approaches, such as making use of rule-based drug discovery systems to sophisticated ML and molecular computing (see Figure 2.1 in Section 2.7). This part (Figure 2.2) will cover methods of computation of individual ADMET parameters, such as absorption (intestinal permeability, solubility, and bioavailability), distribution (blood-brain pathway penetration, Vd, plasma–protein binding), metabolism (cytotoxylgyme P450 interactions, metabolic stability, phase I/II reactions), excretion (renal clearance, bile excretion, and half-life), and toxicity (acute/ chronic toxicity, genotoxicity, cardiotoxicity, and organ toxicity). These forecasts are also based on the theoretical principles of computational chemistry and PK (see Section 2.7), and all these together create the overall guideline to future progress of drug discovery under a very efficient and ethical approach [45].

Figure 2.2 illustrates the interaction of each of the ADMET parameters, which includes Absorption, Distribution, Metabolism, Excretion, and Toxicity parameters in the human body. It shows how drugs are absorbed intestinally and then transported throughout the body via the bloodstream; they are metabolized in the liver using CYP450 enzymes and excreted in the kidneys and bile, as well as the possible toxicity of the drug to organs vital to the body such as the heart and liver. The figure is used to point out the in silico models (QSAR, docking, and PBPK) that replicate these processes in the body to optimize the design of drugs and predict drug safety.

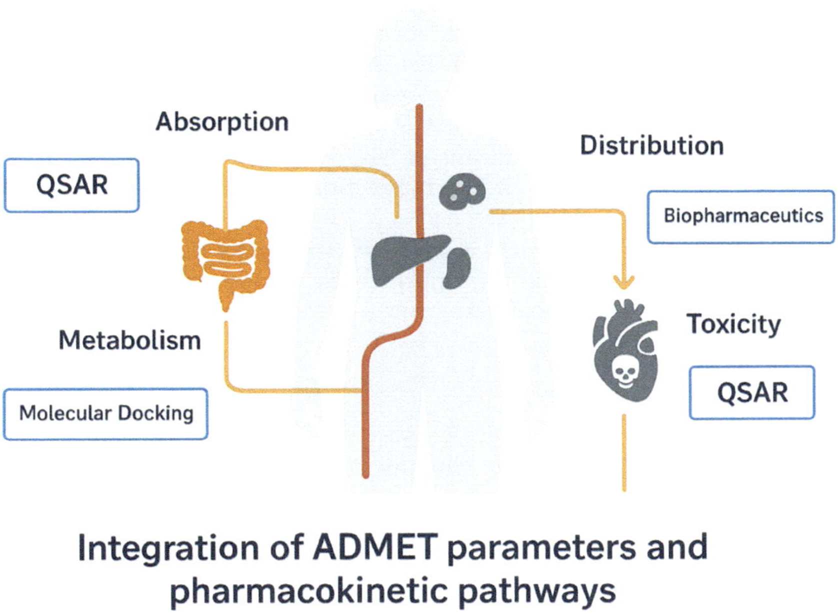

Figure 2.2 Integration of ADMET parameters with human pharmacokinetic pathways.

2.9.1 Absorption

Absorption: This is the point where a drug is absorbed by the blood circulatory system, and this is where drugs are absorbed through the gastrointestinal tract, as with oral drugs. Prediction of first-order parameters (i.e., intestinal permeability, solubility, bioavailability) of molecular determinants (see Table 2.2 in Section 2.7) upon a computational scheme (see Table 2.1 in Section 2.7) enables the early assessment of a drug candidate's absorption profile and facilitates the selection of compounds with favorable pharmacokinetic properties.

Intestinal Permeability: The permeability of the intestine defines how a drug will penetrate the gut epithelia. Structural in silico models are given as follows:

Rule-Based Approaches: The thorough rules suggested by Lipinski, the Rule of Five, and those suggested by Veber, Rules (in the previous Section 2.22) include predictions of permeability depending on MW (MW<500 Da), LogP (<5), TPSA (<140 A2), and rotatable bonds (<10).

QSAR Models: QSAR models have been correlated with data of permeability in vitro through the model of Caco-2 cells (in the likeness of the intestinal epithelium) in terms of descriptors such as the LogP, TSAA, and the number of hydrogen bond donors. Some data mining tools like random forests or support ML (SVMs) are based on ML patterns that especially focus on non-linear relationships to increase accuracy.

PAMPA Simulations: The Parallel Artificial Membrane Permeability Assay (PAMPA) is a model that provides an idea of passive diffusion through artificial membranes. PAMPA in silico is the model that involves crossing the membranes by means of MD simulations, which are carried out through the inference of the interactions between lipophilicity and molecular size.

Transporter-Based Model: Computational models predict interactions with efflux or uptake transporters (e.g., P-glycoprotein or organic anion-transporting polypeptides) using of docking or QSAR to predict transporter affinity [45].

2.9.1.1 Solubility

The aqueous solubility affects the gastrointestinal tract dissolution and absorption. In silico models include the following:

QSPR Models: QSPR models are those that predict solubility based on descriptors such as LogP, MW, and polarizability. Among the ones, the General Solubility Equation can be used to estimate solubility by combining the LogP and melting point. **ML:** Deep learning in the form of deep neural networks (DNNs) or gradient boosters such as XGBoost can predict solubility across a variety of chemical space and use DNNs based on large datasets, such as PubChem or ChEMBL.

Molecular Modeling: It will use MD simulations to model drug–water interactions, giving mechanistic understanding of solubility.

2.9.1.2 Bioavailability

Bioavailability is defined as the proportion of a dose used that enters into the systemic circulation. In silico models are models that synthesize predictions of permeability and solubility with first-pass metabolism.

Rule-Based Filters: Lipinski and Vebers' rules are useful to give the preliminary estimates of bioavailability.

PBPK Modeling: A PBPK model incorporates solubility and permeability data, along with metabolism data, to enable oral bioavailability predictions, taking into consideration such physiological factors as gastric emptying or the intestinal pH.

ML: The bioavailability is forecasted by ML based on the descriptors, such as TPSA, rotatable bonds, or metabolic stability.

Uses and Problems: Software such as SwissADME and ADMETlab are prediction programs that are used to optimize a compound based on absorption parameters. Technical issues include the need to consider active transport processes and physiological heterogeneity (e.g., pH gradients), requiring the use of hybrid in silico/in vitro methods must be used (see Section 2.22) [46].

2.9.2 Distribution

Distribution refers to the process of transportation of a drug to tissues and organs after being absorbed. The BBB penetration, Vd, and plasma–protein binding, which are essential for therapeutic success and safety, are predicted by in silico models.

Penetration of BBB—The penetration of bBB into the central nervous system (CNS) denotes how a drug can file into the CNS, which is relevant to a CNS-targeting drug or avoiding CNS side effects. Models include the following:

Rule-Based Approaches: TPSA [<60 Å^2; LogP >2]: BBB permeability is predicted, whereas highly polar molecules have difficulty crossing the lipid-rich BBB.

QSAR Models: ML models (e.g., SVMs, DNNs) are predictive correlations of descriptors (e.g., LogP, TPSA, and hydrogen bond acceptors) with in vitro model BBB penetration data (e.g., MDCK cells).

Molecular Modeling: MD can be used to study interactions with BBB transporters (e.g. ABC transporters) or lipid bilayers, providing mechanistic understanding.

2.9.2.1 Volume of distribution (Vd)

Vd represents how much of the drug is distributed in the tissues compared to plasma. In silico models include the following:

QSPR Models: Vd correlates with descriptors such as LogP, charge, and MW, which predict Vd and secondary to tissue avidity and plasma affinity.

PBPK Modeling: Reference calculates the distribution of drugs throughout a body compartment (e.g., muscle, fat, liver) but includes physiological activities such as tissue blood flow.

ML: The gradient boosting model used is used to predict Vd of various compounds using a dataset such as DrugBank.

2.9.2.2 Plasma–protein binding

The binding of the drug with plasma proteins, especially albumin or alpha-1-acid glycoprotein, influences the free fraction and effectiveness of the drug. Models include the following:

Molecular Docking: Is an affinity predictor of binding to plasma proteins, involving AutoDock or Glide.

QSAR Models: Find such correlations between the descriptors such as logP and charge and in vitro binding data (equilibrium dialysis).

ML: DNNs are accurate at protein–ligand interactions that are taken into account as complex interactions.

Applications and Challenges: Applications impacting CNS drug design, or the prevention of off-target activity, Some tools such as pkCSM make prediction of distribution parameters. Difficulties in establishing predictions of specific tissue distribution and becoming aware of the inter-individual require combining the analysis with in vitro data.

2.9.3 Metabolism

The transformation of drugs by specific enzymes in the liver mainly affects their activity and clearance. In silico prediction of cytochrome P450 (CYP) interactions and stability in phases I and II.

Sensitivities of cytochrome P450 enzymes—Approximately 75% of medications are metabolized by different types of CYP enzymes (e.g., CYP3A4, CYP2D6). Models include the following:

Molecular Docking: It is used to predict binding affinity with active sites of CYP grouping substrates or inhibitors. CYP-drug interactions are simulated using tools, such as GOLD or the Schrodinger Glide.

QSAR Models: Predict CYP inhibition properties or substrate selectivity based on descriptors such as steric bulk, electronic properties, and LogP. Random forests predict better through ML models.

QM: Networks analyze electronic properties (e.g., HOMO-LUMO maximum) to determine the reactive sites most likely to be subject to CYP metabolism.

2.9.3.1 Metabolic stability

Metabolic stability indicates the ability of a drug to be resistant to enzymatic degradation and throws an impact on half-life as well as the frequency of dosage. Models include the following:

QSPR Models: Models of intrinsic clearance are predicted using descriptors such as MW and lipophilicity that are correlated with in vitro data of microsomal stability.

ML: DNNs and gradient boosting are used to predict stability in any kind of chemical space, with the use of predictive datasets such as ChEMBL.

MD Simulations: Simulate the dynamic interactions among drugs and enzymes in time giving meticulous understanding [45].

2.9.3.2 *Phase I and II reactions*

Phase I Reactions: These are oxidation, reduction, or hydrolysis of the candidate drug, mainly by CYPs. Computer-based tools such as MetaSite or SMARTCyp are able to predict a site of metabolism (SOM) based on electronic and steric properties.

Phase II Reactions: Conjugation, for example, glucuronidation, sulfation by enzymes (UGTs, etc.). QSAR models and docking make predictions of conjugation sites but the information is limited to phase II predictions because of insufficient data.

Applications and Challenges: The use of tools such as BioTransformer forecasts kind pathways that can be optimized structurally to optimize the stability. Hot spots that are tough to solve are rarer metabolic reactions that can no longer rely on phase I/II predictions, as well as the combination of phase I/II predictions with in silico/in vitro to methodologies [47].

2.9.4 Excretion

Excretion refers to the process of getting drugs or metabolites out of the body, which is mainly through the renal or biliary pathways. The in silico models extrapolate renal clearance and biliary excretion, as well as half-life.

Renal Clearance—Renal clearance indicates that the drug is eliminated via the resilience system in the kidneys by filtration, secretion, or reabsorption. Models include the following:

QSPR Models: Forecast clearance with the use of such descriptors as solubility, charge, and MW, which are correlated with the in vitro change in renal clearance data.

PBPK Modeling: The renal excretion is modeled by combining physiological factors (e.g., glomerular filtration rate) and properties of the drug (e.g., with models such as Simcyp).

ML: DNNs or random forests can predict clearance of a wide range of compounds, which is using such datasets as DrugBank [45].

2.9.4.1 *Biliary excretion*

The process of biliary excretion is the feces elimination of drugs through bile to the feces, which is frequently the case with lipophilic drugs. Models include the following:

QSAR Models: Predict biliary clearance based on varying descriptors such as LogP and TPSA, which has got correlations to in vitro hepatocyte data.

Molecular Docking: Biliary–transporter interaction (i. e., of BSEP, MRP2) is estimated, which predicts excretion.

PBPK Modeling: The approach translates biliary excretion throughout the liver parts, with the activity of transporters taken into account [48].

2.9.4.2 *Half-life estimations*

The half-life shows how much time the example of a drug takes to reduce plasma concentration with half. Models include the following:

QSPR models: They predict half-life based on the use of the descriptors such as the rate of clearance and Vd, which were correlated with in vivo data.

PBPK Modeling: PBPK modeling combines an absorption, distribution, and metabolism base to determine the half-life and considers the variability of physiological differences.

ML: DNNs are used to rapidly predict the half-life of complex compounds, enhancing the prediction accuracy of all datasets.

Applications and Challenges: Excretion tools such as ADMET Predictor can be used to estimate excretion parameters, which are useful in dose regimen design. However, this is challenged by difficulties in predicting excretion with transporters and inter-individual differences, which need to be combined with in vitro data [49].

2.9.5 Toxicity

One of the critical areas where precision in toxicity forecasting is essential is with regard to drug safety, which encompasses both acute/chronic toxicity, genotoxicity, cardiotoxicity, and organ toxicity (e.g., hepatotoxicity, nephrotoxicity).

2.9.5.1 *Acute and chronic toxicity*

The following is used to predict acute toxicity (short-term exposure effects) and chronic toxicity (long-term exposure effects):

1. **QSAR Models**: Relate descriptors such as LogP, Log, and MW, and electronic properties with toxicity data, such as in vitro (e.g., LD50 tests) or in vivo data. Random forest and SVMs can improve accuracy in predicting toxicity.
2. **ML**: Predicting toxic ranges (oral LD50) with DNNs using large curated datasets such as Tox21.

3. **Rule-Based Systems**: 3/75 Rule (LogP>3, TPSA<75 A2) Pfizer has a very high toxicity risk compound rule, 3/75 rule [50].

2.9.5.2 Genotoxicity

Genotoxicity refers to a threat of the destruction of DNA and is a serious regulatory issue (see Table 2.4 not other in Section 2.7). Models include the following:

1. **Genotoxicity (i.e., Ames test results):** Prediction depends on molecular properties such as electrophilicity and aromaticity. The ICH M7 guidelines recommend combining rule-based QSAR models, such as Derek Nexus, with statistical models.
2. **Molecular Modeling**: QM computations of reactive metabolites can form DNA adducts.
3. **Consensus Modeling**: Involves the use of different QSAR models together to enhance the ability to predict genotoxicity according to ICH M7.

An advantage of smokeless nicotine gum compared to conventional cigarettes is that smokeless gum won't cause heart problems like high blood pressure [51].

2.9.5.3 Cardiotoxicity (hERG inhibition)

QT prolongation and arrhythmias may occur when one inhibits the hERG potassium channel. Models include the following:

1. **Molecular Docking**: Molecular docking predicts the binding affinity of hERG, where software such as AutoDock models how molecules interact with the channel pore region.
2. **QSAR models**: They correlate descriptors such as lipophilicity and positive charge to hERG data from patch-clamp experiments.
3. **ML**: DNNs and GNNs can be effectively used to predict hERG inhibition based on datasets such as ChEMBL [52].

2.9.5.4 Organ-specific toxicities

1. **Hepatotoxicity**: Forecasted by QSAR models and ML (i.e., by comparing descriptors such as LogP and reactive groups with liver toxicity predictions (i.e., by HepG2 assays). Interactions with liver transporters (e.g., BSEP) are identified by the use of molecular modeling.

2. **Nephrotoxicity**: QSAR and PBPK models determine renal toxicity by considering descriptors such as solubility and clearance, which are correlated with in vitro renal cell data.
3. **Other Toxicities** In other important cases, neurotoxicity, pulmonary toxicity, or dermal toxicity are modeled in the same style but due to the paucity of data, this kind of model is less accurate [53].

Tools—Use and Interest—Whenever an application requires a consistent chemistry infrastructure to comply with regulatory rules (e.g., ICH M7, FDA MIDD), toxicity prediction (using ToxCast and DeepTox) is utilized. Challenges may include anticipating idiosyncratic toxicities, dealing with sparse data needed for predictions on uncommon endpoints, and combining multi-organ toxicity predictions, which may require hybrid approaches. The forecasting of separate ADMET parameters, including ADMET, takes advantage of a wide range of computational optimizations, including rule-set systems, QSAR/QSPR predictions, molecular modeling, or ML (see Tables 2.1 and 2.3 in Section 2.7). These technologies enable accurate early-phase analysis of therapeutic opportunities, ideal drug kinetics and drug safety. These methods meet regulatory expectations (see Table 2.4 in Section 2.7). Absorption models are used to predict permeability, solubility, and bioavailability. Distribution models are used to determine BBB self-distribution, Vd, and plasma–protein binding. Metabolism models identify interactions that facilitate CYP, stability, and excretion models estimate clearance and half-life. Toxicity models determine acute, chronic, and organ effects. In silico ADMET prediction contributes to the increase in efficiency of drug discovery, cost-cutting, and ethical drug development practices, making it an extremely important tool in today's pharmacology. This tool integrates with in vitro data and systems biology (see Section 2.22).

2.10 DATABASE, SOFTWARE, EQUIPMENT, AND CHECKLIST FOR IN SILICO ADMET PREDICTION

In silico ADMET prediction In silico ADMET utilizes a solid tool and software ecosystem to simulate the PK and safety actions of medication candidates. These resources allow researchers to apply computational strategies (see Section 2.22) and predict individual ADMET parameters (see Section 2.7). This helps streamline drug discovery experiments and meet regulatory standards (see Table 2.4 in Section 2.7). The tools, software, and databases used in the field of in silico ADMET prediction are discussed in this section (e.g., open-source tools such as SwissADME, pkCSM, RDKit, commercial software such as Schrodinger Suite, Discovery Studio, MOE, and important databases such as PubChem, ChEMBL, ToxCast and DrugBank). It also addresses workflow systems (e.g., KNIME, Pipeline Pilot), integrating with HTS and big data analytics, and validation strategies, such as measures (e.g., accuracy, precision, recall, ROC-AUC), techniques of cross-validation, applicability domain, quantification of uncertainty, case studies and limitations to predicting novel chemical spaces. Some of these elements when combined, create an effective framework upon which to build up in silico ADMET prediction uses in contemporary pharmacology [54].

2.10.1 Tools and software

In silico ADMET prediction has made use of several open-source and commercial computer tools, each presenting unique features to model molecular properties and ADMET parameters (see Table 2.1 in Section 2.7).

2.10.1.1 Open-source tools

The free open-source software is available, used by many people, and is suitable for academic and preliminary research.

SwissADME: An action logic web-based application that predicts the ADMET parameters, such as the biomagnifiability (using Lipinski's rules of solubility and Veber's p450 temporal interaction rules), solubility, and interaction activities with cytochrome P450. It employs QSAR/QSPR models (see Table 2.3 in Section 2.7) and combines the molecular descriptors such as LogP and TPSA (see Table 2.2 in Section 2.7).

1. **pkCSM**: Predicts pharmacokinetic and toxicology endpoint data (e.g., intestinal permeability, hERG inhibition, and renal clearance) using ML models (e.g., random forests, SVMs). It is helpful in HTS and easy to use for non-experts.
2. **RDKit**: Cheminformatics toolkit that considers molecular descriptors is able to create 2D/3D representations of the molecule, and can perform similarity searches. RDKit has been embedded in our prediction pipelines for ADMET.
3. **OpenBabel**: Software used to convert molecular formats and compute descriptors used in the development of QSAR models, and also in virtual screening.
4. **ChemDes**: Molecular descriptors and fingerprints computational platform, web-based calculator, nanomol, used within QSAR/QSPR conceptual frameworks in ADMET prediction.

Pros: Open-source tools are free, adjustable, and supported, which can be helpful when academic research and small-scale projects should be undertaken.

Cons: Narrow technical support and less sophisticated features than commercial software [55].

2.10.1.2 Commercial software

There is ERP commercial software with advanced functions, improved support, and integration with experimentation workflows that are common in the industrial environment:

1. **Schrodinger Suite**: A complete suite of molecular modeling, including Maestro as a visualization package, Glide as a docking and Jaguar as a QM modeling package. It uses both docking simulations and MD simulations to predict parameters of ADMET such as BBBs penetration and metabolic stability.
2. **Discovery Studio (BIOVIA)**: Parser tools are used to model QSAR, perform molecular docking and PBPK offerings to allow predictions of toxicity (e.g., hepatotoxicity) and pharmacokinetic properties (e.g., bioavailability).
3. **MOE (Molecular Operating Environment)**: Provides QSAR/QSPR, docking, and pharmacophore modeling functions that are used for predicting CYP interactions, solubility, and toxicity. The integrated database in MOE helps in predicting ligand-based ADMET.
4. **ADMET Predictor (Simulations Plus)**: An ML-based predictor of ADMET parameters, such as solubility, permeability, and hERG inhibition, based on ML and PBPK.
5. **StarDrop**: It targets drug-likeness optimization and ADMET optimization and runs DM on QSAR models that involve visualization for next-generation bioavailability and toxicity prediction with visualization tools [56,57].

2.10.2 Most important training and validation databases

Experimental data required to train and validate in silico ADMET models can be found in databases, ensuring that the model is accurate and generalized (refer to Table 2.3 in the theoretical findings section for the type of model).

1. **PubChem**: This is a free database containing millions of chemical structures and biological activity data, including solubility, LogP, and toxic endpoints. PubChem facilitates the expansion of the belief model and virtual grid screening.
2. **ChEMBL**: A manually reviewed collection of data on bioactive molecules, including site-selected AADMET contents such as inhibition at the CYP 4, bioavailability, and hERG binding. ML model development is common in ChEMBL.
3. **ToxCast/Tox21**: These EPA toxicology databases contain high-throughput in vitro toxicity data of thousands of compounds (e.g. hERG inhibition, genotoxicity) used to provide a basis for toxicity models created with QSARs.
4. **DrugBank**: This is an extensive database of approved drugs with their pharmacokinetic data, such as clearance, half-life, and plasma–protein binding, which can be used to train PBPK and QSAR models.
5. **ZINC**: This is a database of commercially available compounds that support virtual screening and calculation of descriptors to be used in predicting ADMET.
6. **HSDB (Hazardous Substances Data Bank)**: Hub of toxicity data relating to environmental and pharmaceutical compounds, can be used to predict toxicities on specific organs.

Applications and Challenges: These databases provide approaches and allow for building extensive model training due to the availability of different chemical and biological data. Some of the challenges include the quality of data (e.g., variations in experimental procedures), the lack of data, and incomplete coverage of new chemical spaces, which would require data curation and pre-processing [58].

2.10.3 Workflow platforms and workflows pipes

This enables workflow platforms to automate ADMET prediction using tools, databases, and computational methods.

1. **KNIME**: An open-source system for processing data analytics and cheminformatics, which includes tools such as RDKit, ChemAxon, and ML copyright lists (e.g., scikit-learn). KNIME aids in the development of QSAR models, calculation of descriptors, and HTS.
2. **Pipeline Pilot (BIOVIA)**: Commercial pipeline modeling platform. A commercial application pipeline modeling platform that combines

QSAR, docking, and PBPK models with databases such as ChEMBL. It assists workflows that are in compliance with regulations.

3. **TIBCO Spotfire**: An analytics platform and data visualization tool that can combine ADMET predictions with experimental data for use in decision-making in drug optimization.
4. **AutoML Platforms**: Other platforms such as H2O.ai or Google Cloud AutoML use ML modeling development to predict ADMET and do not require manual tuning.

Applications and Challenges: Workflow systems improve speed and repeatability, allowing for the wide integration of data sources. Many challenges are encountered, such as the need for computing skills to create pipelines based on specific requirements and the potential incompatibility of tools that need to be utilized to be compatible with another one [59].

2.10.4 Knowledge integration with big data analytics and high-throughput screening opportunity

To process large chemical libraries and complex data, in silico prediction of ADMET is starting to combine with HTS and the big data analytics. HTS software and laboratory programs, such as SwissADME and pkCSM, handle thousands of compounds at a time for ADMET properties, ranking them for consideration in in vitro studies. Virtual HTS is a method that avoids compounds filtered by bioavailability or toxicity in a virtual library through the use of QSAR models and molecular docking. ML models, such as DNNs and GNNs, utilize large datasets, including PubChem, ChEMBL, and ToxCast datasets to make ADMET predictions. Techniques like transfer learning and multi-task learning can be useful for making predictions in sparse data.

Cloud Computing: Services such as AWS or Google Cloud allow predictions of ADMET in a scalable manner to support the computation of large-scale MD and ML models training.

Combination with In Vitro Data: Hybrid models are being used with in silico predictions together with HTS data (e.g. Caco-2 permeability, hERG assays) to enhance modeling, as discussed in Section 2.22 [60].

Applications and Challenges: The integration of HTS and big data is being improved to increase throughput and accuracy to facilitate quick drug discovery. Desirable features include the ability to deal with data heterogeneity, scale computations, and ensure the validity of predictions for new compounds. To meet these restrictions, scientific techniques to tackle them involve transfer learning, multi-task learning, and hybrid in vitro in silico techniques to broaden model coverage and make better predictions in novel chemical space. The drug discovery system, with its ecosystem of tools, software, databases, and validation strategies for in silico prediction of ADMET, is potent. Characteristics such as SwissADME and pkCSM open-source software, as well as entries into the world of commercial software like Schrodinger Suite or MOE, and databases such as PubChem and ChEMBL, enable robust thermodynamic (ADMET) modeling. Predictions are streamlined through workflow systems, such as KNIME and Pipeline Pilot and scaled through integration with large-scale screening and big data analytics. The reliability of the models can be ensured with the help of validation using metrics (e.g., ROC-AUC), cross-validation, analysis of the applicability domain, and the quantification of uncertainty. Case studies provide a comparison of the data collected in silico and through experiments, supporting the idea. Although predicting the ADMET properties of new chemical spaces remains an uphill task, the current research on ML, hybrid predictive methods, and regulatory alignment (see Table 2.4 in Section 2.7 of this paper) frames in silico ADMET prediction as a paradigm shift in pharmacology today, making the merging of inventions and counseling of ethical responsibility both crucial and efficient [61].

2.11 DRUG DISCOVERY AND FUTURE DIRECTIONS

In silico ADMET prediction has now become a part of drug discovery, where researchers can test a drug's pharmacokinetic and safety profiles early in the development cycle. In silico ADMET prediction can address drug candidate selection optimization, eliminate attrition rates, and scientifically contribute to the escalation of regulatory criteria (see Table 2.4 in Section 2.7 of concept) and employs computational approaches, tools, and databases (see also Section 2.62). This chapter addresses the uses of in silico ADMET prediction related to drug discovery, such as early-stage virtual screening, rough optometry, polypharmacology, individualized medicine, population PK, and case studies of successful in silico preclinical predictions in approved drugs. It also comments on the combination with other in silico methods including target identification and pharmacophore modeling. Moreover, it discusses issues such as the lack of data, the problem of complex molecules, and ethical issues and describes the future, including the development of deep learning, quantum computing, multi-omics fusion, and AI-aided design of drugs in real time. Such observations highlight the transformational nature of in silico ADMET forecasting in pharmacology today [62]. **Applications in Drug Discovery:** These applications may be represented using either linear or nonlinear models, depending on the generic definition. ADMET prediction in silico is used at different drug discovery phases, improving efficiency, reducing costs, and promoting ethical progression through the replacement of animal testing. Virtual Screening at a Young Age—Virtually screening Virtual screening relies on in silico ADMET models to screen large libraries of compounds and single compounds based on their favorable pharmacokinetic and safety profiles. Key applications include the following:

HTS: Different tools such as SwissADME and pkCSM (found in Section 2.62) for computing absorption, bioavailability and toxicity evaluating compounds based on rules of thumb (e.g., the rule of five) and QSAR models; and ranking compounds based on their trial conductibility.

Mocking Non-Drug-Like Compounds: Molecular descriptors (i.e. LogP, TPSA; see Table 2.2 in Section 2.7) are used to filter out compounds with poor solubility or high toxicity potential and minimize downstream failure.

Priority on Synthesis: Synthetic chemistry activities are directly optimized with in silico predictions to find compounds with good ADMET characteristics, saving time and resources.

A case in point: Using pkCSM virtual screening of 10,000 compounds, 500 compounds were identified (with oral bioavailability and low hERG inhibition), reducing the experimentation workload by 95% [63].

2.11.1 Lead optimization

The in silico prediction of virtual ADMET is used in conjunction with structural changes to optimize the pharmacokinetic and safety profiles of a lead compound:

1. **Optimizing Absorption**: Table 2.1 (see Section 2.7) provides QSAR models and MD simulations that could be used to optimize of solubility or intestinal permeability. This optimization may involve reducing TPSA or making changes in LogP.
2. **Reducing Toxicity**: Molecular docking and QSAR are aiding in identifying structural characteristics associated with toxicity (e.g., hERG binding) and can be used to make structural modifications to reduce risks.
3. **Assisting in Metabolic Stability**: SMARTCyp is available to predict sites of cytochrome P450 metabolism, allowing for the adjustment of half-life.

Example: ADMET Predictor Lead optimization of a kinase inhibitor minimized hERG inhibition by altering a cationic moiety increasing cardiotoxicity safety [64].

2.11.2 Naked innovations existing drugs and polypharmacology

In silico prediction in ADMET facilitates drug repurposing (validation of improved uses of known drugs) and polypharmacology (designing drugs with multiple targets):

1. **Drug Repurposing**: Predicting Drug Discovery Drugs that are already in use may have appropriate PK to help us reach the desired target. For example, QSAR modeling can be used to examine the bioavailability and toxicity of antivirals used as anticancer agents.
2. **Polypharmacology**: There are in silico tools to predict the off-target and ADMET properties of multi-target drugs, which aim to strike a balance between their efficacy and safety. Using techniques such as molecular docking or ML (e.g., DNNs), compounds with promising multi-target activity can be identified.

Example: In silico ADMET prediction revealed good bioavailability and low toxicity of remdesivir in the repurposing of Ebola for COVID-19 treatment, which was confirmed by clinical trials [65].

2.11.3 Individualized pharmacokinetics and population pharmacokinetics

In silico ADMET prediction assists in personalized medicine and population PK: Inter-individual variability is considered:

1. **Personalized**: The CYP polymorphism Personalists Personalized models combine both genetic and physiological information (e.g., CYP polymorphisms) to predict the ADMET behavior of certain patient populations to optimize dosing schedules.
2. **Population PK**: PBPK simulations are models of drug behavior in a population (e.g., pediatrics, elderly) based on physiological aspects of population data such as organ size and blood flow, enhanced with physiological databases such as DrugBank.
3. **Pharmacogenomics**: AI involves estimating the association of genetic differences (e.g., CYP2D6 poor metabolizers) with metabolism to create individualized treatment [66].

Scenario: Simcyp (PBPK modeling) enabled warfarin to be given to patients with CYP2C9 variants at the optimum dose, resulting in better therapeutic outcomes.

Case studies: Why predictions in approved drugs worked

1 SOFOSBUVIR (ANTIVIRAL):

In Silico Contributions: QSAR and PBPK models indicated high oral bioavailability and low hepatotoxicity which led to the development of hepatitis C treatment.

Result: The positive ADMET profile of sofosbuvir was confirmed through clinical trials and therefore approved by the FDA in 2013.

Impact: The prioritization of one lead compound led to reduced time in the development and reduced cost of the development [67].

2 IMATINIB (ANTICANCER):

In Silico Contribution: Both molecular docking and QSAR models showed a clear indication of strong kinase binding, with low hERG efficacy, maximizing the selection of hERG cardiotoxicity.

Result: This confirmed the safety and effectiveness of imatinib in the treatment of chronic myeloid leukemia and it was approved in 2001.

Impact: Provided the efficacy of in silico ADMET as part of drug development in oncology therapy [68].

3 OSELTAMIVIR (ANTIVIRAL):

In Silico Contribution: The QSAR models and MD simulations were used to predict high intestinal permeability and metabolic stability, enabling further use as an oral treatment for influenza.

Deliverable: Predictions were confirmed by clinical trials, and approved in 1999.

Impact: A higher rate of accelerated development due to selective dependence on promising analogs [69].

4 APIXABAN (ANTICOAGULANT):

In Silico Contribution: The consensus modeling (QSAR and docking) model projected low CYP3A4 and good bioavailability, which reduces drug–drug interactions.

Outcome: Stroke prevention is approved and has a tolerable safety profile (2012).

Impact: Declines in the use of animal testing through the use of in silico predictions [70].

2.11.4 Combination with other in silico methodologies

Computational techniques, software, and computer-aided drug discovery in in silico ADMET prediction are commonly used together with other predictive approaches such as target identification, molecular docking, and pharmacophore modeling to discover target drugs (e.g., GPCRs, kinases). These techniques can be used in conjunction with ADMET models to predict good PK. For example, the Schrodinger Suite takes into account docking and ADMET predictions. Pharmacophore Modeling: This determines the essential molecular features that bind to the target and is applied in combination with QSAR models to optimize ADMET. OED frameworks are assisted using tools such as MOE. Virtual Screening: ADMET predictions are coupled with virtual ligand-based (QSAR) and structure-based (docking) screening to select hits with optimal efficacy and safety. MD simulations: simulate drug–target and drug–membrane interactions, which are complimentary to ADMET predictions on distribution and absorption (see Section 2.7).

Case Study: Pharmacophore modeling integration of SwissADME found a dual kinase inhibitor that was optimally bioavailable and non-toxic, with in vitro assays confirming the results [71].

2.12 CHALLENGES AND FUTURE DIRECTIONS

Analyzing its achievements, in silico ADMET prediction seems to have to contend with challenges that diminish its usefulness, accompanied by emergent changes that ensure its future impact.

Issues of Data Scarcity, Quality, and Training Set Bias:

Data Scarcity: Sparsity of data on complex endpoints (e.g., idiosyncratic toxicities, phase II metabolism) is a problem affecting the accuracy of the model. Currently, databases such as ToxCast and ChEMBL (see Section 2.62), do not adequately cover new scaffolds.

Quality Problems: Changes in experimental protocols (e.g., variable assay conditions) introduce noise, compromising model reliability. For example, discrepancies in data from hERG assays can lead to false positives.

One-Sided Training Sets: By focusing more on models based on known scaffolds and less on novel ones, bias influences models in both directions, resulting in reduced performance with novel compounds in any model.

Solutions: Build high-quality datasets, use transfer learning with similar data, and combine multi-omics data to increase coverage [71].

2.12.1 Handling complex molecules

1. **Biologics**: (e.g., monoclonal antibodies) proteins and peptides have a complex form; descriptor calculation and QSAR modeling becomes difficult. Present instruments are affordable to small molecules.
2. **Nanomaterials**: Nanoparticles applied in drugs delivery do not have standardized AB, which constrain the ability of ADMET predictions.
3. **Natural Products and Macrocycles**: They possess unusual physicochemical characteristics that are frequently not within the scope of existing models.

Solutions: Name biologics with special descriptors, simulate nanomaterials using coarse-grained MD simulations, and train on a wider collection of various molecular classes [72].

2.12.2 AI model ethical and interpretability concerns

1. **Black-Box Issues**: Deep learning models (e.g., DNNs, GNNs) cannot be understood, which makes them challenging to accept by the regulation organizations (see Table 2.4 in Section 2.7). As an example, it is important to make decisions about predicting the toxicity of a compound.

Ethical Implications: In training data (e.g. lack of representation of some populations) may result in unequal pre-predictions, especially in personalized medicine.

Regulatory Compliance: Regulatory agencies such as the FDA and EMA demand the use of transparent models that pose a challenge to the application of music of complex AI systems.

The solutions include interpretable models (e.g., SHAP to estimate the role of features) to get an evaluation of uncertainty or to use conformal prediction and to make sure that data sets are diverse and represent the whole population [73].

2.12.3 Emerging trends

Deep Learning Improvements: GNNs and transformers are deep learning concepts that enhance the prediction of ADMET by considering long-range interactions and molecular graph models. Multi-task learning forecasts several ADMET endpoints at a time, making it more efficient.

Quantum computing enables quantum algorithms to run quickly on QM calculations and MD simulations, allowing for the highest-precision ADMET predictions of complex molecules.

Multi-Omics Integration: Integrating genomics, proteomics, and metabolomic fundamentals with ADMET models achieves the personalized application of medicine by predicting unique response of the patient.

Cloud-Based Platforms: AWS and Google cloud will allow making ADMET predictions on a crisis basis, which will feed on laptops of virtual screening and real-time analytics [74].

2.12.4 Prospects for real-time forecasting and association with AI-Aided design

1. **Real-Time Predictions**: AI platforms on clouds and automated pipelines (e.g., KNIME, Pipeline Pilot) can be used to perform real-time predictions of ADMET to facilitate fast decision-making in drug discovery.
2. **AI-Assisted Drug Design**: Generative AI models (generative adversarial networks, variational autoencoders and others) are developed to improve the design of novel molecules with optimal ADMET characteristics using QSAR or PBPK models.

Case Example: GNNs were used to optimize a bioavailability parameter and minimize hERG inhibition of an antiviral compound with the assistance of AI, reducing the lead optimization time by 50% [75].

The most recent developments will be in deep learning, quantum computing and integration of multi-omics, all contributing to increased accuracy of prediction and extension to complex molecules. The design will be facilitated by AI in real time and should be in line with regulatory and ethical objectives (e.g., 3Rs principles). In silico ADMET of drugs is a radical instrument in drug discovery and is used in virtual screening, lead optimization, drug repurposing, polypharmacology, and personalized medicine. Examples of cases making its case include sofosbuvir and imatinib, which show how it has been successful in helping the approved drugs. It is complemented by further in silico techniques, including integration and target identification, pharmacophore modeling and databases such as ChEMBL. Nevertheless, limitations such as data unavailability, management of multifaceted molecules, and AI interpretability should be overcome to achieve the greatest benefits. New tendencies in deep learning, quantum computers, and integration of multi-omics and the AI-related design in real time make ADMET prediction radicalize and become a fundamental staple of effective, ethically novel, and personalized pharmacology of the contemporary pharmacology [76].

2.13 CONCLUSION

In silico ADMET intent prediction has revolutionized drug discovery by enabling efficient and ethical practices in pharmacokinetic and safety profile evaluation through early assessments using data measurements. The complex ADMET predictions have been examined within the framework of this chapter, including the current state of theoretical basis, computational methods to address the problem, tools, validation, issues, and future perspectives of ADMET. Several essential lessons will emphasize the transformative nature of drug development in the present day, as well as the future of medicine involving pharmacology. In conclusion, personalized nanomedicine offers significant potential to improve ADMET prediction and toxicity profiling in breast cancer treatment. By leveraging advanced nanocarriers and precision tools like molecular profiling and AI, these systems enable more accurate modeling of drug behavior and safety. Integrated with next-generation trial designs and preclinical models, personalized nanomedicine enhances the prediction of therapeutic outcomes and minimizes toxicity, paving the way for safer, more effective precision oncology strategies [77].

Theoretical underpinnings: The prediction by in silico ADMET models, the fundamental theoretical bases of computational chemistry (e.g., QM, MD: see Table 2.1 in Section 2.7), PK, and molecular descriptors (e.g., LogP, TPSA: see Table 2.2) come together to offer strong predictive frameworks through which the QSPR/QSAR models are pronounced (see Table 2.3). The principles allow the prediction of complicated biological events, including intestinal permeability and genotoxicity, as explained in the section where the prediction of individual ADMET parameters is elaborated. Computational approaches, such as rule-based approaches (e.g., the Rule of Five), ML (e.g., deep neural networks), and molecular modeling provide computational methods that are useful in providing accurate predictions. Ensemble approaches and hybrid methods increasing their reliability (see Section 2.22). By using SwissADME, pkCSM, and Schrodinger Suite, which are facilitated by PubChem and ChEMBL, databases that assist in HTS and validation, one can ensure compliance with regulatory theories such as those pertaining to FDA, EMA, and ICH (see Table 2.4 in Section 2.7). In silico ADMET prediction is used in drug discovery to facilitate early-stage virtual screening, lead optimization, drug repurposing, polypharmacology, and personalized medicine. Case studies of sofosbuvir and imatinib show that the approach has succeeded in guiding the approval of drugs based on predictions of bioavailability, toxicity, and metabolic stability, thereby reducing the development time and costs. It can also be used in combination with other methods, such as pharmacophore modeling and target identification. As seen in Section 2.74, it is even more useful.

Nonetheless, despite these developments, in silico ADMET prediction has pitfalls related to the lack of data, biases during training datasets as well as the inability to model complex molecules, such as biologics or nanomaterials. Legal and ethical issues, specifically, the interpretability of AI models, are hindrances toward the acceptance of regulation and fair predictions. The solution to these is the interdisciplinary cooperation between computational chemists, pharmacologists, data scientists, regulatory experts to globalize the data maintenance, measurements of validation, and reporting guidelines. These activities will provide reproducibility, reliability, and functionality following guidelines such as ICH M7.

In the future, the predominant trends of emerging fields such as deep learning (e.g., GNNs), quantum computing, and integration of multi-omics will boost predictive performance and enable applications in new chemical space. Simplified AI-assisted drug development, cloud-based applications, and real-time design will make it even faster as the development of cloud-based applications and real-time drug development will also speed up design and provide the rapid development and optimization. It is possible to improve the field by enhancing interdisciplinary interaction and uniformity to combat the existing constraints, providing safer and more effective therapeutics readily. In conclusion, ADMET in silico prediction can be regarded as one of the pillars of recent drug discovery, as it presents an effective and ethical method to facilitate drug development and tailor therapy. This chapter provides the tools that enable a researcher to utilize them and advance future pharmacology based on data.

REFERENCES

1. Walker DK. The Use of Pharmacokinetic and Pharmacodynamic Data in the Assessment of Drug Safety in Early Drug Development. *Br J Clin Pharmacol.* 2004 Dec;58(6):601–608. doi: 10.1111/j.1365-2125.2004.02194.x.
2. Cheng F, Li W, Liu G, Tang Y. In Silico ADMET Prediction: Recent Advances, Current Challenges and Future Trends. *Curr Top Med Chem.* 2013;13(11):1273–1289. doi: 10.2174/1568026611313999003 3.
3. Alagga AA, Pellegrini MV, Gupta V. Drug Absorption. [Updated 2024 Feb 27]. In: StatPearls [Internet]. Treasure Island (FL): StatPearls Publishing; 2025 Jan. Available from: https://www.ncbi.nlm.nih.gov/books/NBK557405/.
4. Dulsat J, López-Nieto B, Estrada-Tejedor R, Borrell JI. Evaluation of Free Online ADMET Tools for Academic or Small Biotech Environments. *Molecules.* 2023 Jan 12;28(2):776. doi: 10.3390/molecules28020776.
5. Sun D, Gao W, Hu H, Zhou S. Why 90% of Clinical Drug Development Fails and How to Improve It? *Acta Pharm Sin B.* 2022;12:3049–3062. doi: 10.1016/j.apsb.2022.02.002.
6. Hubrecht RC, Carter E. The 3Rs and Humane Experimental Technique: Implementing Change. *Animals (Basel).* 2019 Sep 30;9(10):754. doi: 10.3390/ani9100754.

7. Daoud NE, Borah P, Deb PK, Venugopala KN, Hourani W, Alzweiri M, Bardaweel SK, Tiwari V. ADMET Profiling in Drug Discovery and Development: Perspectives of In Silico, In Vitro and Integrated Approaches. *Curr Drug Metab.* 2021;22(7):503–522. doi: 10.2174/1389200222666210705122913.
8. Ferreira LG, Dos Santos RN, Oliva G, Andricopulo AD. Molecular Docking and Structure-Based Drug Design Strategies. *Molecules.* 2015 Jul 22;20(7):13384–13421. doi: 10.3390/molecules200713384.
9. Benet LZ, Hosey CM, Ursu O, Oprea TI. BDDCS, the Rule of 5 and druggability. *Adv Drug Deliv Rev.* 2016 Jun 1;101:89–98. doi: 10.1016/j.addr.2016.05.007.
10. The International Transporter Consortium. Giacomini KM, Huang SM, Tweedie DJ, Benet LZ, Brouwer KL, Chu X, Dahlin A, Evers R, Fischer V, Hillgren KM, Hoffmaster KA, Ishikawa T, Keppler D, Kim RB, Lee CA, Niemi M, Polli JW, Sugiyama Y, Swaan PW, Ware JA, Wright SH, Wah Yee S, Zamek-Gliszczynski MJ, Zhang L. Membrane Transporters in Drug Development. *Nat Rev Drug Discov.* 2010;9:215–236. doi: 10.1038/nrd3028.
11. Lipinski CA, Lombardo F, Dominy BW, Feeney PJ. Experimental and Computational Approaches to Estimate Solubility and Permeability in Drug Discovery and Development Settings. *Adv Drug Deliv Rev.* 2001;46:3–26. doi: 10.1016/s0169–409x(00)00129-0.
12. Lipinski CA. Drug-Like Properties and the Causes of Poor Solubility and Poor Permeability. *J Pharmacology Toxicol Methods.* 2000;44:235–249. doi: 10.1016/s1056–8719(00)00107-6.
13. Komura H, Watanabe R, Mizuguchi K. The Trends and Future Prospective of In Silico Models from the Viewpoint of ADME Evaluation in Drug Discovery. *Pharmaceutics.* 2023 Nov 12;15(11):2619. doi: 10.3390/pharmaceutics15112619.
14. Jia L, Gao H. Machine Learning for In Silico ADMET Prediction. *Methods Mol Biol.* 2022;2390:447–460. doi: 10.1007/978-1-0716–1787–8_20.
15. Tzeliou CE, Mermigki MA, Tzeli D. Review on the QM/MM Methodologies and Their Application to Metalloproteins. *Molecules.* 2022 Apr 20;27(9):2660. doi: 10.3390/molecules27092660.
16. Warshel A, Levitt M. Theoretical Studies of Enzymic Reactions: Dielectric, Electrostatic and Steric Stabilization of the Carbonium Ion in the Reaction of Lysozyme. *J Mol Biol.* 1976;103:227–249. doi: 10.1016/0022–2836(76)90311-9.
17. Grogan S, Preuss CV. Pharmacokinetics. [Updated 2023 Jul 30]. In: StatPearls [Internet]. Treasure Island (FL): StatPearls Publishing; 2025 Jan-. Available from: https://www.ncbi.nlm.nih.gov/books/NBK557744/
18. Wu F, Zhou Y, Li L, Shen X, Chen G, Wang X, Liang X, Tan M, Huang Z. Computational Approaches in Preclinical Studies on Drug Discovery and Development. *Front Chem.* 2020 Sep 11;8:726. doi: 10.3389/fchem.2020.00726.
19. Veber DF, Johnson SR, Cheng HY, Smith BR, Ward KW, Kopple KD. Molecular Properties That Influence the Oral Bioavailability of Drug Candidates. *J Med Chem.* 2002 Jun 6;45(12):2615–2623. doi: 10.1021/jm020017n.

20. Boczar D, Michalska K. A Review of Machine Learning and QSAR/QSPR Predictions for Complexes of Organic Molecules with Cyclodextrins. *Molecules.* 2024 Jul 2;29(13):3159. doi: 10.3390/molecules29133159.
21. Boczar D., Michalska K. Cyclodextrin Inclusion Complexes with Antibiotics and Antibacterial Agents as Drug-Delivery Systems - A Pharmaceutical Perspective. *Pharmaceutics.* 2022;14:1389. doi: 10.3390/pharmaceutics14071389.
22. Daoui O, Elkhattabi S, Chtita S, Elkhalabi R, Zgou H, Benjelloun AT. QSAR, Molecular Docking and ADMET Properties In Silico Studies of Novel 4,5,6,7-Tetrahydrobenzo[D]-Thiazol-2-Yl Derivatives Derived from Dimedone as Potent Anti-tumor Agents Through Inhibition of C-Met Receptor Tyrosine Kinase. *Heliyon.* 2021 Jul 3;7(7):e07463. doi: 10.1016/j.heliyon.2021.e07463.
23. Iyengar R, Zhao S, Chung SW, Mager DE, Gallo JM. Merging Systems Biology with Pharmacodynamics. *Sci Transl Med.* 2012 Mar 21;4(126):126ps7. doi: 10.1126/scitranslmed.3003563.
24. Tsouka S, Masoodi M. Metabolic Pathway Analysis: Advantages and Pitfalls for the Functional Interpretation of Metabolomics and Lipidomics Data. *Biomolecules.* 2023 Jan 27;13(2):244. doi: 10.3390/biom13020244.
25. Niu J, Straubinger RM, Mager DE. Pharmacodynamic Drug-Drug Interactions. *Clin Pharmacol Ther.* 2019 Jun;105(6):1395–1406. doi: 10.1002/cpt.1434.
26. Bassan A, Alves VM, Amberg A, Anger LT, Auerbach S, Beilke L, Bender A, Cronin MTD, Cross KP, Hsieh JH, Greene N, Kemper R, Kim MT, Mumtaz M, Noeske T, Pavan M, Pletz J, Russo DP, Sabnis Y, Schaefer M, Szabo DT, Valentin JP, Wichard J, Williams D, Woolley D, Zwickl C, Myatt GJ. *In Silico* Approaches in Organ Toxicity Hazard Assessment: Current Status and Future Needs in Predicting Liver Toxicity. *Comput Toxicol.* 2021 Nov;20:100187. doi: 10.1016/j.comtox.2021.100187.
27. Madabushi R, Seo P, Zhao L, Tegenge M, Zhu H. Review: Role of Model-Informed Drug Development Approaches in the Lifecycle of Drug Development and Regulatory Decision-Making. *Pharm Res.* 2022 Aug;39(8):1669–1680. doi: 10.1007/s11095-022-03288-w.
28. Paul P, Colin PJ, Musuamba Tshinanu F, Versantvoort C, Manolis E, Blake K. Current Use of Physiologically Based Pharmacokinetic modeling in New Medicinal Product Approvals at EMA. *Clin Pharmacol Ther.* 2025 Mar;117(3):808–817. doi: 10.1002/cpt.3525.
29. Hasselgren C, Bercu J, Cayley A, Cross K, Glowienke S, Kruhlak N, Muster W, Nicolette J, Reddy MV, Saiakhov R, Dobo K. Management of Pharmaceutical ICH M7 (Q)SAR Predictions - The Impact of Model Updates. *Regul Toxicol Pharmacol.* 2020 Dec;118:104807. doi: 10.1016/j.yrtph.2020.104807.
30. Viceconti M, Pappalardo F, Rodriguez B, Horner M, Bischoff J, Musuamba Tshinanu F. *In Silico* Trials: Verification, Validation and Uncertainty Quantification of Predictive Models Used in the Regulatory Evaluation of Biomedical Products. *Methods.* 2021 Jan;185:120–127. doi: 10.1016/j.ymeth.2020.01.011.

31. Zhang Y, Luo M, Wu P, Wu S, Lee TY, Bai C. Application of Computational Biology and Artificial Intelligence in Drug Design. *Int J Mol Sci.* 2022 Nov 5;23(21):13568. doi: 10.3390/ijms232113568.
32. Wang S, Donga G, Shenga C. Structural Simplification: An Efficient Strategy in Lead Optimization. *Acta Pharm Sin B.* 2019;9:880–901. doi: 10.1016/j.apsb.2019.05.004.
33. Fattouche M, Belaidi S, Abchir O, Al-Shaar W, Younes K, Al-Mogren MM, Chtita S, Soualmia F, Hochlaf M. ANN-QSAR, Molecular Docking, ADMET Predictions, and Molecular Dynamics Studies of Isothiazole Derivatives to Design New and Selective Inhibitors of HCV Polymerase NS5B. *Pharmaceuticals (Basel).* 2024 Dec 18;17(12):1712. doi: 10.3390/ph17121712.
34. Aloui M, Er-rajy M, Imtara H, Goudzal A, Zarougui S, El fadili M, Arthur DE, Mothana RA, Noman OM, Tarayrah M, et al. QSAR Modelling, Molecular Docking, Molecular Dynamic and ADMET Prediction of Pyrrolopyrimidine Derivatives as Novel Bruton's Tyrosine Kinase (BTK) Inhibitors. *Saudi Pharm J.* 2024;32:101911. doi: 10.1016/j.jsps.2023.101911.
35. Khamouli S., Belaidi S., Bakhouch M., Chtita S., Hashmi M.A., Qais F.A. QSAR Modeling, Molecular Docking, ADMET Prediction and Molecular Dynamics Simulations of Some 6-Arylquinazolin-4-Amine Derivatives as DYRK1A Inhibitors. *J Mol Struct.* 2022;1258:132659. doi: 10.1016/j.molstruc.2022.132659.
36. Venkataraman M, Rao GC, Madavareddi JK, Maddi SR. Leveraging Machine Learning Models in Evaluating ADMET Properties for Drug Discovery and Development. *ADMET DMPK.* 2025 Jun 7;13(3):2772. doi: 10.5599/admet.2772.
37. Wang D, Liu W, Shen Z, Jiang L, Wang J, Li S, Li H. Deep Learning Based Drug Metabolites Prediction. *Fron Pharmacol.* 2020;10:1586. https://doi.org/10.3389/fphar.2019.01586 10.3389/fphar.2019.01586
38. Ryu JY, Lee JH, Lee BH, Song JS, Ahn S, Oh KS. PredMS: A Random Forest Model for Predicting Metabolic Stability of Drug Candidates in Human Liver Microsomes. *Bioinformatics.* 2022;38:364–368.
39. Mostafa F, Howle V, Chen M. Machine Learning to Predict Drug-Induced Liver Injury and Its Validation on Failed Drug Candidates in Development. *Toxics.* 2024;12(12):385. doi: 10.3390/toxics12060385.
40. Bunker A, Róg T. Mechanistic Understanding from Molecular Dynamics Simulation in Pharmaceutical Research 1: Drug Delivery. *Front Mol Biosci.* 2020 Nov 25;7:604770. doi: 10.3389/fmolb.2020.604770.
41. Almulihi A, Saleh H, Hussien AM, Mostafa S, El-Sappagh S, Alnowaiser K, Ali AA, Refaat Hassan M. Ensemble Learning Based on Hybrid Deep Learning Model for Heart Disease Early Prediction. *Diagnostics (Basel).* 2022 Dec 18;12(12):3215. doi: 10.3390/diagnostics12123215.
42. Absar M, Zaidah AR, Mahmood A, Ahmad S, Ejaz H, Ahmed N, Nik Hashim NHH, Yean CY. A Review of In Silico and In Vitro Approaches in the Fight Against Carbapenem-Resistant Enterobacterales. *J Clin Lab Anal.* 2025 May;39(9):e70018. doi: 10.1002/jcla.70018.
43. Chou P, Shannar A, Pan Y, Dave PD, Xu J, Kong AT. Application of Physiologically-Based Pharmacokinetic (PBPK) Model in Drug Development and in Dietary Phytochemicals. *Curr Pharmacol Rep.* 2025;11(1):45. doi: 10.1007/s40495-025-00427-w.

44. Ezuruike U, Zhang M, Pansari A, De Sousa Mendes M, Pan X, Neuhoff S, Gardner I.. Guide to Development of Compound Files for PBPK Modeling in the Simcyp Population-Based Simulator. *CPT Pharmacometrics Syst Pharmacol.* 2022;11(7):805–821.
45. Jung W, Goo S, Hwang T, Lee H, Kim YK, Chae JW, Yun HY, Jung S. Absorption Distribution Metabolism Excretion and Toxicity Property Prediction Utilizing a Pre-Trained Natural Language Processing Model and Its Applications in Early-Stage Drug Development. *Pharmaceuticals (Basel).* 2024 Mar 17;17(3):382. doi: 10.3390/ph17030382.
46. Price G, Patel DA. Drug Bioavailability. [Updated 2023 Jul 30]. In: StatPearls [Internet]. Treasure Island (FL): StatPearls Publishing; 2025 Jan-. Available from: https://www.ncbi.nlm.nih.gov/books/NBK557852/.
47. Phang-Lyn S, Llerena VA. Biochemistry, Biotransformation. [Updated 2023 Aug 14]. In: StatPearls [Internet]. Treasure Island (FL): StatPearls Publishing; 2025 Jan-. Available from: https://www.ncbi.nlm.nih.gov/books/NBK544353/.
48. Sharifi M, Ghafourian T. Estimation of Biliary Excretion of Foreign Compounds Using Properties of Molecular Structure. *AAPS J.* 2014 Jan;16(1):65–78. doi: 10.1208/s12248-013-9541-z.
49. Hallare J, Gerriets V. Elimination Half-Life of Drugs. [Updated 2025 May 3]. In: StatPearls [Internet]. Treasure Island (FL): StatPearls Publishing; 2025 Jan. Available from: https://www.ncbi.nlm.nih.gov/books/NBK554498/.
50. Zhang R, Wen H, Lin Z, Li B, Zhou X. Artificial Intelligence-Driven Drug Toxicity Prediction: Advances, Challenges, and Future Directions. *Toxics.* 2025 Jun 23;13(7):525. doi: 10.3390/toxics13070525.
51. Menz J, Götz ME, Gündel U, Gürtler R, Herrmann K, Hessel-Pras S, Kneuer C, Kolrep F, Nitzsche D, Pabel U, Sachse B, Schmeisser S, Schumacher DM, Schwerdtle T, Tralau T, Zellmer S, Schäfer B. Genotoxicity Assessment: Opportunities, Challenges and Perspectives for Quantitative Evaluations of Dose-Response Data. *Arch Toxicol.* 2023 Sep;97(9):2303–2328. doi: 10.1007/s00204-023-03553-w.
52. Creanza TM, Delre P, Ancona N, Lentini G, Saviano M, Mangiatordi GF. Structure-Based Prediction of hERG-Related Cardiotoxicity: A Benchmark Study. *J Chem Inf Model.* 2021 Sep 27;61(9):4758–4770. doi: 10.1021/acs.jcim.1c00744.
53. Low Y, Uehara T, Minowa Y, Yamada H, Ohno Y, Urushidani T, Sedykh A, Muratov E, Kuz'min V, Fourches D, Zhu H, Rusyn I, Tropsha A. Predicting Drug-Induced Hepatotoxicity Using QSAR and Toxicogenomics Approaches. *Chem Res Toxicol.* 2011 Aug 15;24(8):1251–1262. doi: 10.1021/tx200148a.
54. Kar S, Roy K, Leszczynski J. *In Silico* Tools and Software to Predict ADMET of New Drug Candidates. *Methods Mol Biol.* 2022;2425:85–115. doi: 10.1007/978-1-0716–1960–5_4.
55. Kar S, Leszczynski J. Open Access *In Silico* Tools to Predict the ADMET Profiling of Drug Candidates. *Expert Opin Drug Discov.* 2020 Dec;15(12):1473–1487. doi: 10.1080/17460441.2020.1798926.
56. Aminpour M, Montemagno C, Tuszynski JA. An Overview of Molecular Modeling for Drug Discovery with Specific Illustrative Examples of Applications. *Molecules.* 2019 Apr 30;24(9):1693. doi: 10.3390/molecules24091693.

57. Agu PC, Afiukwa CA, Orji OU, Ezeh EM, Ofoke IH, Ogbu CO, Ugwuja EI, Aja PM. Molecular Docking as a Tool for the Discovery of Molecular Targets of Nutraceuticals in Diseases Management. *Sci Rep.* 2023 Aug 17;13(1):13398. doi: 10.1038/s41598-023-40160-2.
58. Chang Y, Hawkins BA, Du JJ, Groundwater PW, Hibbs DE, Lai F. A Guide to In Silico Drug Design. *Pharmaceutics.* 2022 Dec 23;15(1):49. doi: 10.3390/pharmaceutics15010049.
59. Leonis G, Melagraki G, Afantitis A. Open Source Chemoinformatics Software Including KNIME Analytics. In J. Leszczynski, A. Kaczmarek-Kedziera, T.G. Puzyn, M. Papadopoulos, H.K. Reis, & M. Shukla (Eds.), *Handbook of Computational Chemistry.* 2017 Dec 9:2201–30. Springer, Cham. doi: 10.1007/978-3-319-27282-5_57.
60. Szymański P, Markowicz M, Mikiciuk-Olasik E. Adaptation of High-Throughput Screening in Drug Discovery-Toxicological Screening Tests. *Int J Mol Sci.* 2012;13(1):427–452. doi: 10.3390/ijms13010427.
61. Ayon NJ. High-Throughput Screening of Natural Product and Synthetic Molecule Libraries for Antibacterial Drug Discovery. *Metabolites.* 2023 May 2;13(5):625. doi: 10.3390/metabo13050625.
62. Mobley D.L., Guthrie J.P. FreeSolv: A Database of Experimental and Calculated Hydration Free Energies, with Input Files. [(accessed on 1 March 2024)]; *J Comput Aided Mol Des.* 2014;28:711–720. doi: 10.1007/s10822-014-9747-x.
63. Vamathevan J, Clark D, Czodrowski P, Dunham I, Ferran E, Lee G, Li B, Madabhushi A, Shah P, Spitzer M, Zhao S. Applications of Machine Learning in Drug Discovery and Development. *Nat Rev Drug Discov.* 2019 Jun;18(6):463–477. doi: 10.1038/s41573-019-0024–5.
64. Opo FADM, Rahman MM, Ahammad F, Ahmed I, Bhuiyan MA, Asiri AM. Structure Based Pharmacophore Modeling, Virtual Screening, Molecular Docking and ADMET Approaches for Identification of Natural Anti-cancer Agents Targeting XIAP Protein. *Sci Rep.* 2021;11:4049. doi: 10.1038/s41598-021-83626-x.
65. Cheng F. In Silico Oncology Drug Repositioning and Polypharmacology. *Methods Mol Biol.* 2019;1878:243–261. doi: 10.1007/978-1-4939–8868-6_15.
66. Boobis A, Gundert-Remy U, Kremers P, Macheras P, Pelkonen O. In Silico Prediction of ADME and Pharmacokinetics. Report of an Expert Meeting Organised by COST B15. *Eur J Pharm Sci.* 2002 Dec;17(4–5):183–193. doi: 10.1016/s0928–0987(02)00185-9.
67. Xia A, Liu J, Tang C, Xu H. A Case Report of Sofosbuvir Induced Drug Eruption and Literature Review. *IDCases.* 2025 May 9;40:e02249. doi: 10.1016/j.idcr.2025.e02249.
68. Gagic Z, Ruzic D, Djokovic N, Djikic T, Nikolic K. *In silico* Methods for Design of Kinase Inhibitors as Anticancer Drugs. *Front Chem.* 2020 Jan 8;7:873. doi: 10.3389/fchem.2019.00873.
69. Davies BE. Pharmacokinetics of Oseltamivir: An Oral Antiviral for the Treatment and Prophylaxis of Influenza in Diverse Populations. *J Antimicrob Chemother.* 2010 Apr;65(Suppl 2):ii5–ii10. doi: 10.1093/jac/dkq015.
70. Agrawal A, Manna B. Apixaban. [Updated 2024 Feb 22]. In: StatPearls [Internet]. Treasure Island (FL): StatPearls Publishing; 2025 Jan-. Available from: https://www.ncbi.nlm.nih.gov/books/NBK507910/.

71. Agamah FE, Mazandu GK, Hassan R, Bope CD, Thomford NE, Ghansah A, Chimusa ER. Computational/In Silico Methods in Drug Target and Lead Prediction. *Brief Bioinform.* 2020 Sep 25;21(5):1663–1675. doi: 10.1093/bib/bbz103.
72. Han X, Shih J, Lin Y, Chai Q, Cramer SM. Development of QSAR Models for *In Silico* Screening of Antibody Solubility. *MAbs.* 2022 Jan-Dec;14(1):2062807. doi: 10.1080/19420862.2022.2062807.
73. Sheu Y-h. Illuminating the Black Box: Interpreting Deep Neural Network Models for Psychiatric Research. *Front Psychiatry.* 2020;11:551299. doi: 10.3389/fpsyt.2020.551299.
74. Reiser P, Neubert M, Eberhard A, Torresi L, Zhou C, Shao C, Metni H, van Hoesel C, Schopmans H, Sommer T, Friederich P. Graph Neural Networks for Materials Science and Chemistry. *Commun Mater.* 2022;3:93. doi: 10.1038/s43246-022-00315-6.
75. Djoumbou-Feunang Y, Wilmot J, Kinney J, Chanda P, Yu P, Sader A, Sharifi M, Smith S, Ou J, Hu J, Shipp E, Tomandl D, Kumpatla SP. Cheminformatics and Artificial Intelligence for Accelerating Agrochemical Discovery. *Front Chem.* 2023 Nov 29;11:1292027. doi: 10.3389/fchem.2023.1292027.
76. Sutanto H, Fetarayani D. Integrating Artificial Intelligence into Small Molecule Development for Precision Cancer Immunomodulation Therapy. *NPJ Drug Discov.* 2025;2:25. doi: 10.1038/s44386-025-00029-y.
77. Sahana HD, Dasegowda M, Basalingappa KM, Vishwas S. Future Perspectives-Personalized Nanomedicine and Next-Generation Clinical Trials in Breast Cancer. In R. Pal, B. Ghosh, M. Koli, & P. Dutta (Eds.), *Next-Gen Nanomedicine for Breast Cancer: From Bench to Bedside and Beyond* (pp. 243–273). Deep Science Publishing. 2025. doi: 10.70593/978-93-7185-537-2_10.

Chapter 3

Multi-target and polypharmacology-oriented drug design

Koushik Sen[1], *Bhanumati Sarkar*[2], *Tanmay Sanyal*[3], *Suman Adhikari*[4], *and Nithar Ranjan Madhu*[5*]

[1]Department of Zoology, Jhargram Raj College, Jhargram, West Bengal, India

[2]Department of Botany, Acharya Prafulla Chandra College, West Bengal, India

[3]Department of Zoology, Krishnagar Govt College, West Bengal, India

[4]Department of Chemistry, Govt. Degree College, Dharmanagar, Tripura, India

[5]Department of Zoology, Acharya Prafulla Chandra College, West Bengal, India

[*]Corresponding Author: nitharranjan.madhu@gmail.com

3.1 INTRODUCTION TO MULTI-TARGET DRUG DESIGN

Over recent decades, pharmaceutical research has undergone a marked conceptual and methodological evolution. The historically dominant "one drug–one target" doctrine, which focused on highly selective agents acting on a single protein or pathway, proved effective for a range of monogenic diseases and many infectious conditions in which clinical manifestations could be directly traced to dysfunction of a specific molecular entity. As our understanding of disease biology has progressed, particularly through systems-level analyses, it has become evident that the most prevalent disorders, such as malignancies, neurodegenerative diseases, metabolic syndrome, and long-standing inflammatory conditions, are intrinsically polygenic and multifactorial. These pathologies involve extensive disturbance of interlinked, often redundant signaling networks rather than isolated pathway defects (Boran et al., 2010; Tang et al., 2014; Roy et al., 2024).

This realization has driven a strategic transition toward therapeutic concepts that engage multiple targets concurrently. Multi-target drug design seeks to exploit the networked nature of biological regulation by

DOI: 10.1201/9781003672760-3

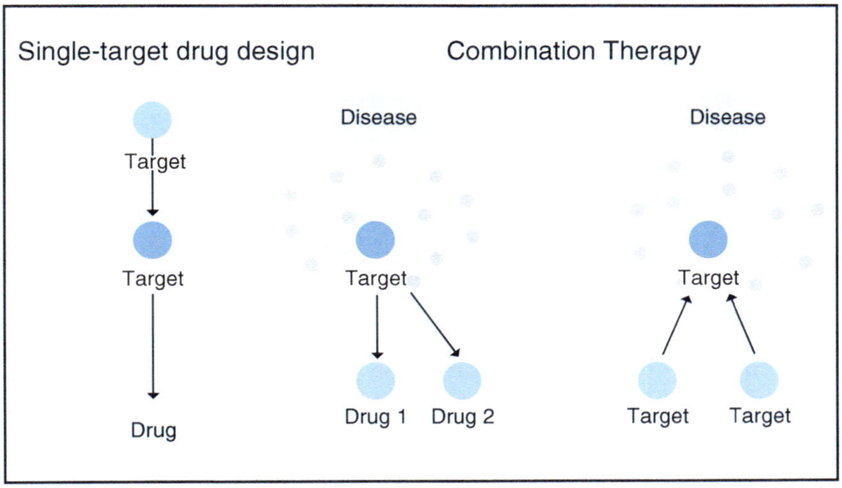

Figure 3.1 Evolution from single-target to multi-target drug design.

deliberately modulating several disease-relevant molecular nodes in an orchestrated manner, thereby achieving more robust and durable clinical responses. The aim is to counteract the compensatory network rewiring and bypass circuitry that are frequently responsible for drug resistance and loss of efficacy with single-target therapies (Anighoro et al., 2014).

Within this framework, a multi-target agent is conceived as a rationally optimized single molecule—or, in some contexts, a carefully coordinated pharmacological regimen—that produces concerted modulation of two or more distinct targets central to disease pathogenesis. Rather than relying on stepwise or sequential targeting, as is typical in empiric combination therapy with independent drugs, multi-target design aims to induce distributed, network-level rebalancing of dysregulated systems. In this way, the intrinsic robustness and plasticity of cellular networks are addressed at a systems level rather than by focal inhibition of a single node.

The scientific justification for multi-target drug discovery stems from contributions in network pharmacology, systems biology, and quantitative modeling of drug–network interactions. These disciplines collectively illustrate how biological systems integrate and respond to simultaneous perturbations, and how optimal combinations of target engagements can maximize therapeutic benefit while constraining the emergence of resistance and dose-limiting toxicities (Hopkins, 2008; Zhao and Iyengar, 2012). Consequently, multi-target design has become a central strategy for tackling complex diseases that have proven refractory to classical monotherapeutic approaches (Figure 3.1).

3.2 CONCEPT AND PRINCIPLES OF POLYPHARMACOLOGY

Polypharmacology refers to both a pharmacodynamic phenomenon and a deliberate design paradigm in which a single small molecule engages multiple, structurally and functionally distinct biological targets within regulatory networks (Reddy et al., 2013). This definition encompasses unintended, empirically observed target promiscuity—often revealed through post-approval safety monitoring or broad screening panels—as well as intentionally engineered multi-target profiles created during the hit-to-lead and lead optimization stages.

The foundational concept underlying polypharmacology is rooted in the organization of protein–ligand interaction space. Many drug-like molecules display measurable affinities (commonly with IC_{50} or Kd values in the nano- to low micromolar range) toward multiple members of structurally related protein families, such as kinase domains sharing conserved ATP-binding motifs, orthosteric or allosteric pockets on G-protein-coupled receptors (GPCRs), or enzymes with homologous cofactor-binding regions (Boran et al., 2010). Families such as the human kinome (~500 kinases), the extensive GPCR repertoire (~800 receptors with shared seven-helix architecture), and the large serine protease superfamily provide a structural basis that facilitates multi-target recognition by relatively compact small molecules.

It is important to distinguish between the two principal forms of multi-target behavior. On the one hand, target promiscuity can arise as a byproduct of certain physicochemical characteristics of the ligand, such as elevated lipophilicity, high aromatic content, broad hydrogen-bonding capabilities, or increased molecular size. On the other hand, carefully engineered polypharmacology is achieved through intentional medicinal chemistry interventions that fine-tune molecular structure to deliver synergistic activity against a preselected set of targets. Quantitative analyses have demonstrated relationships between global molecular descriptors—especially lipophilicity (e.g., logP, TPSA), size, and basicity—and a compound's likelihood of binding to multiple off-targets when screened across large protein panels (Luo et al., 2013).

Within drug–target interaction networks, aminergic GPCRs—including dopamine, serotonin, and adrenergic receptors—frequently appear as "hub" nodes, reflecting both their evolutionary conservation and the structural similarity of their endogenous ligands. These properties contribute to their unusually high susceptibility to engagement by diverse small molecules (Tang et al., 2014; Reddy et al., 2013). In contrast, systems pharmacology-driven approaches emphasize the design of multi-target profiles that are mechanistically aligned with disease networks rather than merely tolerated as incidental promiscuity. In this rational context, an "optimal" polypharmacological agent is one that achieves a favorable balance between desired multi-target efficacy and minimal off-target toxicity, leading to a superior overall therapeutic index compared with uncontrolled promiscuity (Table 3.1).

Table 3.1 Protein families prone to polypharmacology and example ligands

Protein family	*Key structural features*	*Example multi-target ligands*	*Primary therapeutic applications*
Kinases (~500 members)	Conserved ATP-binding cleft	Sunitinib (VEGFR/PDGFR/c-Kit)	Oncology (angiogenesis inhibition)
GPCRs (~800 members)	Seven-transmembrane helices	Aripiprazole (D2/5-HT1A)	Neuropsychiatry (schizophrenia)
Serine Proteases	Catalytic triad (Ser-His-Asp)	Rivaroxaban (Factor Xa/IXa)	Thrombosis prevention
Nuclear Receptors	Ligand-binding domain helix	Bezafibrate (PPARα/γ/δ)	Dyslipidemia

3.3 RATIONALE FOR MULTI-TARGET DRUG DISCOVERY

The strategic motivation for multi-target drug discovery is grounded in several interrelated empirical and theoretical observations derived from network pharmacology and systems biology. A central insight is that many pathological systems—especially tumors, pathogenic microorganisms, and chronically inflamed tissues—possess an inherent capacity to adapt to selective perturbation of individual nodes. This adaptation often occurs through compensatory pathway activation and signaling bypass mechanisms that restore network functionality despite ongoing pharmacological inhibition.

A well-characterized example arises in non-small-cell lung cancer (NSCLC) treated with epidermal growth factor receptor (EGFR) tyrosine kinase inhibitors such as erlotinib and gefitinib. Prolonged exposure commonly leads to resistant subclones harboring diverse genetic modifications, including the EGFR T790M gatekeeper mutation, MET amplification that activates alternative survival signaling axes, and NRAS or RAF1 alterations that enhance downstream MEK–ERK activity. These distinct changes converge functionally to circumvent the blockade of the original EGFR target (Ramirez et al., 2016). Multi-target strategies that simultaneously interfere with EGFR and key bypass routes—such as concurrent inhibition of EGFR and MET or dual targeting of EGFR and MEK—can significantly postpone or even prevent the emergence of resistant populations, thereby improving both progression-free and overall survival (Cheng et al., 2019).

A second major rationale arises from the multifaceted nature of complex diseases. Disorders such as Alzheimer's disease exemplify pathologies in which multiple, intertwined processes—including amyloid-β (Aβ) accumulation, tau hyperphosphorylation, microglial and astroglial activation, oxidative stress, and mitochondrial dysfunction—cooperate to drive neurodegeneration. Clinical experience has repeatedly demonstrated that agents targeting only a single pathological hallmark (e.g., Aβ or tau alone) rarely

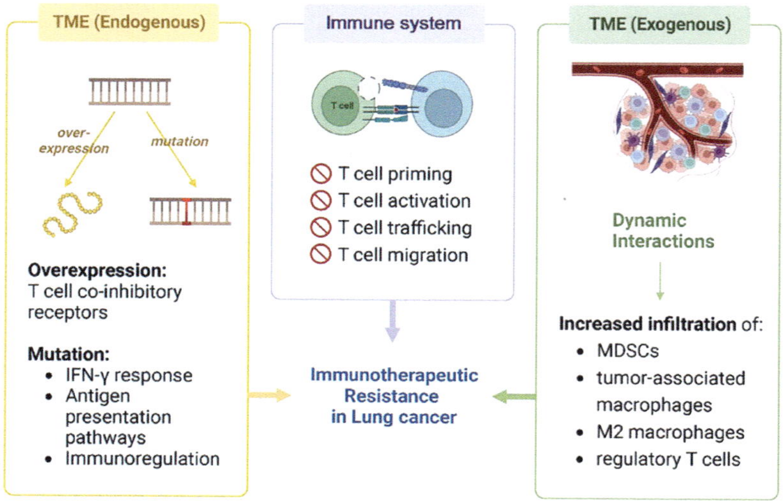

Figure 3.2 Resistance mechanisms in NSCLC (non-small cell lung cancer) and the need for multi-target strategies (Kahraman et al., 2024).

produce meaningful, sustained benefits in late-stage trials (Van der Schyf et al., 2011; Chopade et al., 2022). Multi-target approaches that address several pathogenic mechanisms in parallel are thus conceptually more suited to altering the overall disease trajectory.

Third, from a pharmacokinetic–pharmacodynamic perspective, single-molecule multi-target agents offer practical and clinical advantages over regimens employing multiple drugs. A single active entity simplifies dosing, improves adherence, and yields more predictable absorption, distribution, metabolism, and elimination (ADME) characteristics. This reduction in regimen complexity is particularly valuable in elderly patients or those with multimorbidity requiring polypharmacy. Moreover, single-entity multi-target drugs intrinsically limit the scope for drug–drug interactions mediated through hepatic enzymes or transporters, a frequent concern when combining several independent agents (Ryszkiewicz et al., 2023). These considerations collectively provide a compelling rationale for prioritizing multi-target design, especially in therapy-resistant, heterogeneous, and multifactorial diseases (Figure 3.2).

3.4 MULTI-TARGET DRUGS VS COMBINATION THERAPY

Although multi-target therapy and combination therapy both aim to engage more than one disease-relevant mechanism, they differ conceptually and operationally. Conventional combination therapy typically involves the simultaneous administration of two or more independently developed drugs, each possessing its own pharmacokinetic profile, safety liabilities, and regulatory history. Examples include highly active antiretroviral therapy (HAART) regimens, which integrate nucleoside and non-nucleoside reverse transcriptase inhibitors with protease inhibitors, and oncology protocols that pair cytotoxic chemotherapeutics with targeted kinase inhibitors. Such strategies permit flexible choice of agents and enable relatively rapid translation of promising combinations into clinical use when individual components are already available (Paoli et al., 2024).

However, combination therapy brings substantial complexity. Each agent contributes additional off-target interactions and toxicity risks, creating a cumulative burden that can be difficult to manage. Pharmacokinetic interactions—mediated through overlapping metabolic pathways or transporter systems—may produce unpredictable changes in exposure, complicating dose optimization. Patients may also struggle with adherence to multi-drug regimens, especially when dosing schedules are intricate or pill burdens are high. Furthermore, understanding the mechanistic contributions of each drug to overall efficacy and toxicity can be challenging given the intertwined PK/PD interactions.

Rationally designed multi-target drugs follow a different strategy. Here, the set of intended targets is selected prospectively, guided by systems pharmacology and disease network analyses, and then integrated into a single chemical framework. This design philosophy offers several important advantages. First, the pharmacokinetic behavior is governed by a single ADMET profile, enabling more straightforward modeling and clinical management. Second, off-target interactions are confined to one scaffold, which can be systematically optimized to minimize undesirable activities. Third, achieving effective concentrations at multiple targets may be more efficient, as a single molecule, rather than multiple agents, must reach adequate exposure levels. Fourth, regulatory development is consolidated under a single IND/NDA pathway instead of complex combination approval processes (Anighoro et al., 2014; Cheng et al., 2019).

Significant challenges balance these benefits. Designing a molecule that simultaneously satisfies potency, selectivity, and pharmacokinetic requirements for multiple targets is inherently complex. Moreover, opportunities to adjust doses independently across different targets are limited, as modulation of all engaged targets is constrained to the same concentration range. Nevertheless, when executed successfully, multi-target single agents can deliver superior resistance management, more streamlined clinical use, and mechanistically coherent network modulation compared with traditional polypharmacy (Table 3.2).

Table 3.2 Comparison of multi-target drugs vs. combination therapy

Aspect	*Multi-target drugs*	*Combination therapy*
Pharmacokinetics	Single ADME profile; predictable exposure	Multiple profiles; risk of interactions
Adherence	Simplified dosing (e.g., once-daily pill)	Complex regimens; higher non-compliance
Toxicity management	Confined to one scaffold; easier optimization	Cumulative risks; harder to attribute
Development speed	Single IND; faster approval	Separate approvals; longer for combos
Examples	Sunitinib (multi-kinase)	HAART (multiple antiretrovirals)
Key limitation	Fixed target ratios	Flexible but unpredictable synergies

3.5 STRATEGIES FOR MULTI-TARGET DRUG DESIGN

Modern medicinal chemistry and chemical biology have developed a spectrum of strategies for deliberately crafting molecules with multi-target activity profiles. One widely used approach is fragment-based multi-target design. This method starts from the observation that structurally diverse ligands often share pharmacophoric fragments that bind to different targets. By identifying recurring motifs—such as benzimidazole cores in kinase inhibitors, quinolone scaffolds in metalloprotein inhibitors, or indole-based fragments in GPCR ligands—chemists can assemble hybrid molecules that incorporate key elements from multiple parent compounds (Morphy et al., 2006).

Through this fragment-hybridization tactic, dual- or multi-specific molecules have been generated. Ladostigil (TV3326) is an illustrative example that combines structural components derived from acetylcholinesterase inhibitors with propargylamine moieties characteristic of monoamine oxidase B (MAO-B) inhibitors. The resulting compound exhibits both cholinesterase inhibition and MAO-B inhibition, conferring symptomatic and neuroprotective benefits in neurodegenerative models (Van der Schyf et al., 2011). Likewise, memoquin and related derivatives integrate features that enable amyloid-β chelation and acetylcholinesterase inhibition within a polyamine–quinone framework and have shown encouraging activity in transgenic Alzheimer's disease models.

A second strategy is structure-guided multi-target design, which leverages high-resolution structural information from X-ray crystallography, cryo-electron microscopy, and computational docking. Investigation of kinase domain structures across the human kinome, for example, has revealed extensive conservation of the ATP-binding cleft, enabling the development of inhibitors that concurrently target multiple kinases. Sunitinib and sorafenib exemplify this concept: both compounds inhibit a spectrum of receptor tyrosine kinases, including VEGFR-2/3, PDGFR-α/β, c-Kit, and Flt-3, thereby exerting broad antiangiogenic and antitumor effects (Gridelli et al., 2007). Structure-informed design also facilitates the identification of allosteric sites across different proteins that can accommodate a single ligand, potentially enabling multi-target engagement via conformationally compatible binding pockets with reduced energetic penalties.

A third, complementary strategy is phenotype-driven discovery. Rather than beginning with a predefined target, this approach initiates with high-throughput screening against disease-relevant cellular or organismal phenotypes—such as inhibition of tau aggregation, enhancement of Aβ clearance, or suppression of proinflammatory cytokine production. Compounds that restore normal phenotypes are selected, and their molecular targets are subsequently elucidated using transcriptomics, proteomics, affinity-based capture, or genetic perturbation techniques (Tang et al., 2014; Boran et al., 2010). By operating in a target-agnostic manner, this strategy naturally enriches for molecules whose polypharmacology is aligned with complex disease mechanisms rather than constrained by preconceived single-target hypotheses.

An emerging and increasingly influential paradigm is artificial intelligence (AI)-assisted polypharmacological design. Machine learning, generative neural networks, and reinforcement learning are now routinely used to explore large regions of chemical space in search of molecules with predefined multi-target activity profiles. The POLYGON framework is a representative example, using reinforcement learning to generate de novo dual inhibitors of MEK1 and mTOR. Out of 32 synthesized candidates, approximately three-quarters showed quantifiable activity against the intended targets, with off-target profiles comparable to clinically approved kinase inhibitors (Munson et al., 2024).

Graph convolutional neural networks that encode amino acid sequences or structural features of protein targets alongside molecular fingerprints of ligands can predict multi-target binding patterns with increasing accuracy. Deep learning models trained on extensive bioactivity datasets, such as those available for kinases, can now forecast both desired target engagement and the likelihood of interacting with anti-targets implicated in toxicity. This predictive capability allows chemists to steer optimization toward beneficial polypharmacology while avoiding high-risk off-target interactions (Xiong et al., 2021) (Figure 3.3).

3.6 COMPUTATIONAL APPROACHES IN POLYPHARMACOLOGY

Computational technologies constitute a central pillar of multi-target drug design, providing tools for mapping disease networks, identifying promising target combinations, and optimizing ligand structures to achieve desired polypharmacological profiles. Within this domain, network pharmacology

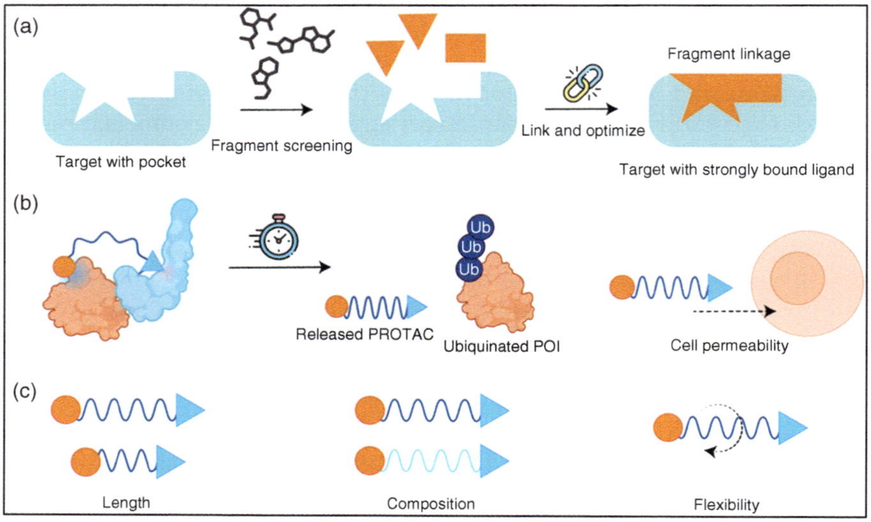

Figure 3.3 Fragment-based hybridization in multi-target design (Gharbi and Mercado, 2024). (a) Fragment-based ligand assembly onto a target binding pocket. (b) Application of fragment-linked hybrids in targeted protein degradation via PROTACs (c) Key linker design parameters governing PROTAC and hybrid molecule performance.

constructs and analyzes graphs representing proteins, genes, and pathways implicated in specific diseases. These networks integrate information from protein–protein interaction repositories (e.g., STRING, BioGRID), genetics (e.g., GWAS datasets), transcriptomics comparing diseased and healthy tissues, and pathway annotation resources such as KEGG and Reactome (Tang et al., 2014; Hopkins, 2008).

Graph-theoretic analyses, using measures such as degree, betweenness, and closeness centrality, as well as community detection algorithms, help identify hub nodes and bottlenecks whose modulation may produce disproportionately large effects on disease network behavior. Network propagation methods and diffusion-based algorithms, exemplified by platforms such as SyndrumNET, assess the proximity and complementarity between drug targets and disease modules. By quantifying how different target sets overlap or complement disease-associated subnetworks, these tools can predict multi-target combinations that maximize therapeutic disruption while minimizing redundancy and off-target impacts (Iida et al., 2024).

Structure-based virtual screening and molecular docking remain indispensable tools for predicting and rationalizing multi-target interactions. Using software such as AutoDock, DOCK, or MOE, candidate ligands are computationally docked into a series of protein structures representing potential targets, and binding scores are calculated using empirical or physics-inspired scoring functions that account for van der Waals contacts, electrostatics, solvation effects, and conformational strain (Hsin et al.,

2013; Agamah et al., 2019). Increasingly, machine learning-enhanced scoring functions trained on experimentally resolved protein–ligand complexes are being incorporated to improve accuracy.

A major difficulty in structure-based polypharmacology is capturing protein flexibility and induced-fit effects, especially for allosteric and transient binding sites. Molecular dynamics simulations, although computationally demanding, can help explore conformational ensembles and reveal dynamic pockets capable of accommodating ligands across multiple targets (Agamah et al., 2019).

Ligand-based virtual screening approaches complement structure-based methods by inferring activity against new targets from similarities to known ligands. Quantitative structure–activity relationship (QSAR) models and classification schemes (based on support vector machines, random forests, or other algorithms) are trained on datasets of compounds with annotated target profiles, enabling predictions for new structures lacking co-crystal data (Sydow et al., 2019). Chemical similarity metrics—including Tanimoto coefficients based on molecular fingerprints (e.g., ECFP4) and more sophisticated graph kernels—are used to prioritize compounds likely to share multi-target activities with reference molecules.

Network-based target prediction represents a unifying layer, combining drug–target interactions with disease gene networks. The "complementary exposure" principle suggests that effective drug combinations tend to involve target sets that collectively map to the disease module but remain sufficiently separated within the interactome to avoid excessive redundancy and overlapping toxicity (Cheng et al., 2019). This framework has been employed to retrospectively rationalize successful clinical combinations and prospectively nominate novel multi-target strategies for experimental validation (Figure 3.4).

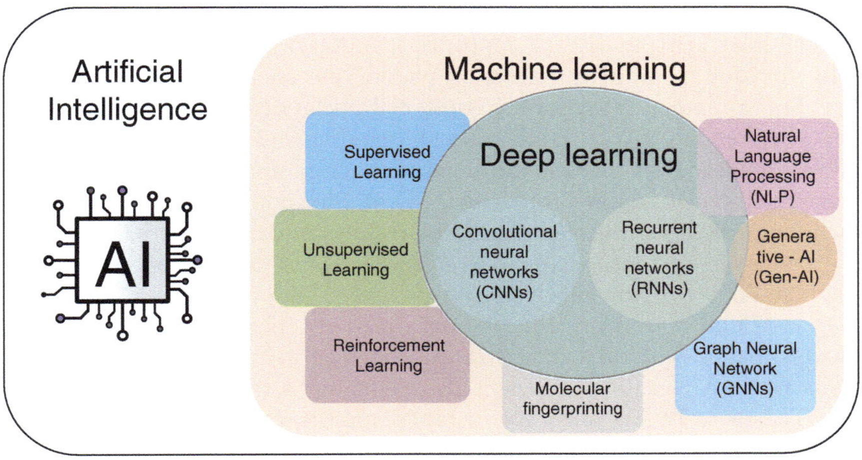

Figure 3.4 Overview of artificial intelligence (AI) and its key components (Kant et al., 2025).

3.7 EXPERIMENTAL VALIDATION TECHNIQUES

The rigorous validation of multi-target compounds demands integration of biochemical, cellular, and in vivo methodologies to demonstrate target engagement, functional pathway modulation, and disease-relevant efficacy, while at the same time defining and limiting off-target effects.

Surface plasmon resonance (SPR) is a widely used biophysical technique for characterizing protein–ligand interactions in real time without the need for labeling. It provides kinetic parameters such as association (ka) and dissociation (kd) rate constants and enables calculation of equilibrium dissociation constants (Kd) across large target panels (Hsin et al., 2013; Shepherd et al., 2022). SPR is particularly valuable for detecting low-affinity fragment interactions (Kd in the 10–100 μM range), evaluating selectivity across related protein families, and triaging fragments or leads suitable for further medicinal chemistry optimization. Newer biosensor platforms, including biolayer interferometry (BLI) and microfluidic-based systems, can extend throughput and support screening against dozens to hundreds of targets with modest material requirements.

Functional kinase assays are central to profiling multi-target kinase inhibitors. These assays typically measure the inhibition of substrate phosphorylation, either by immunodetection methods, luminescence-based ATP consumption readouts, or radiometric phosphate transfer assays. Large-scale kinase panels covering hundreds of human kinases allow quantification of activity profiles, identification of undesired off-target interactions, and ranking of compounds based on potency and selectivity (Xiong et al., 2021). Dose–response studies using concentration series facilitate the determination of IC_{50} values across the intended target set and off-targets, providing a quantitative basis for balancing efficacy against safety.

Cellular phenotypic assays extend validation into disease-relevant biological contexts. Induced pluripotent stem cell (iPSC)-derived neurons, cancer cell lines bearing disease-appropriate mutations, and primary immune cells are used to evaluate effects on phenotypes such as tau or Aβ pathology, inflammatory cytokine secretion, cell proliferation, or apoptosis (Van der Schyf et al., 2011). High-content imaging and multiplexed readouts allow simultaneous assessment of multiple endpoints, capturing the integrated outcome of multi-target engagement. Phosphoproteomic techniques, including mass spectrometry-based workflows and antibody-based arrays, are employed to map pathway-level changes and confirm modulation of predicted downstream substrates.

In vivo studies in transgenic, xenograft, or genetically engineered animal models are critical for demonstrating that in vitro multi-target activities translate into therapeutic benefit. In Alzheimer's disease models such as APP/PS1 or 3xTg-AD mice, for instance, multi-target agents are evaluated for their capacity to reduce Aβ plaques, ameliorate tau pathology, suppress neuroinflammatory markers, and restore cognitive performance in

Table 3.3 Key experimental techniques for multi-target validation

Technique	*Application*	*Throughput*	*Key outputs*
Surface plasmon resonance (SPR)	Binding kinetics across target panels	Medium	ka, kd, Kd
Kinase panel assay	Profiling kinase inhibitors	High	IC_{50}, selectivity scores
High-content imaging	Phenotypic screening in cells	High	Multiplexed readouts
PK/PD modeling	In vivo exposure–response relationships	Low	Target occupancy, efficacy

behavioral assays (Van der Schyf et al., 2011). In oncology, xenograft and genetically engineered models are used to evaluate tumor growth inhibition, survival benefits, and delay in resistance emergence compared with monotherapies. Pharmacokinetic/pharmacodynamic (PK/PD) modeling further helps define relationships between drug exposure, target occupancy, and phenotypic response, supporting rational dose selection and scheduling (Table 3.3).

3.8 APPLICATIONS IN MAJOR DISEASE DOMAINS

3.8.1 Cancer

Oncology has been at the forefront of multi-target drug development, reflecting the complexity and adaptability of tumor signaling networks. Cancer cells often rely on multiple, partially redundant pathways for proliferation, survival, angiogenesis, and metastasis (Madhu, Sarkar, Roychoudhury et al., 2023; Madhu, Biswas, et al., 2024). Single-target therapies can therefore be undermined by alternative signaling routes or clonal evolution leading to resistant subpopulations.

Agents such as sunitinib and sorafenib illustrate the impact of multi-target kinase inhibition. Both compounds inhibit a suite of receptor tyrosine kinases—including PDGFR, VEGFR-1/2/3, c-Kit, and Flt-3—thereby simultaneously targeting tumor cell proliferation and the angiogenic processes necessary for tumor vascularization (Gridelli et al., 2007; Kim et al., 2009). Sunitinib, for example, has been approved for metastatic renal cell carcinoma (RCC) and gastrointestinal stromal tumors (GIST), demonstrating superior objective response rates and progression-free survival compared with interferon-α in RCC (Kim et al., 2009).

In parallel, immuno-oncology has embraced multi-target concepts through combined blockade of immune checkpoints. The pairing of anti-CTLA-4 and anti-PD-1/PD-L1 antibodies removes inhibitory signals at both T cell priming and effector stages, yielding more profound and durable immune responses than single-agent checkpoint inhibition (Wojtukiewicz et al., 2021; He et al., 2020). Regimens such as ipilimumab plus nivolumab in melanoma exemplify this effect, improving overall survival at the expense of increased, but manageable, immune-related toxicities. Other rationally chosen combinations, such as CDK4/6 inhibitors with endocrine therapy in hormone receptor-positive breast cancer or dual RAF/MEK inhibition in BRAF-mutant melanoma, reflect an increasing reliance on systems-level reasoning to select complementary targets.

3.8.2 Neurodegenerative disorders

Neurodegenerative diseases provide classical examples of disorders with multifactorial and evolving pathogenesis, encompassing protein misfolding and aggregation, excitotoxicity, oxidative stress, mitochondrial dysfunction, neuroinflammation, and synaptic failure (Madhu, Sarkar, Roychoudhury et al., 2023; Madhu, Biswas, et al., 2024; Biswas et al., 2024; Das et al., 2024). Consequently, multi-target strategies have gained prominence as more appropriate interventions than single-mechanism agents.

Memoquin and related molecules illustrate a multi-target approach to Alzheimer's disease. Their polyamine–quinone scaffold permits inhibition of Aβ aggregation, direct sequestration of Aβ species, scavenging of reactive oxygen species, and acetylcholinesterase inhibition to enhance cholinergic neurotransmission (Van der Schyf et al., 2011). In transgenic models, these agents have shown robust effects on cognitive performance, amyloid burden, and oxidative damage.

Ladostigil (TV3326) is another archetypal multi-target compound, designed by fusing structural elements conferring acetylcholinesterase inhibition with those responsible for MAO-B inhibition. This dual action supports both symptomatic improvement and neuroprotection by correcting cholinergic deficits and reducing oxidative stress resulting from aberrant monoamine metabolism (Van der Schyf et al., 2011).

Glycogen synthase kinase-3 (GSK-3) has emerged as a particularly attractive hub target in neurodegeneration because it phosphorylates tau, influences amyloid precursor protein processing, and contributes to α-synuclein pathology (Lei et al., 2011). Inhibiting GSK-3—via lithium, small-molecule inhibitors such as CHIR99021 and AR-A014418, or indirectly through Akt pathway activation—simultaneously affects multiple pathogenic substrates, reducing tau hyperphosphorylation, Aβ generation, and α-synuclein accumulation, thereby providing multi-dimensional neuroprotection in preclinical models.

3.8.3 Infectious diseases

Multi-target approaches also underpin modern treatments for chronic infectious diseases, notably HIV infection. HAART regimens combine nucleoside reverse transcriptase inhibitors, non-nucleoside reverse transcriptase inhibitors, and protease inhibitors to disrupt multiple steps of the viral life cycle. This layered blockade markedly reduces the probability of resistance development compared with monotherapy or dual therapy (Eron et al., 2000; Maenza et al., 1998). Triple-drug combinations can achieve sustained suppression of plasma HIV RNA to below detection thresholds in a high proportion of treatment-naïve patients, with durable responses maintained over years (Eron et al., 2000).

More recently, network-based methods such as SyndrumNET have been applied to predict synergistic combinations in other diseases, including chronic myeloid leukemia (CML), by integrating genomic, transcriptomic, interactome, and disease association data (Iida et al., 2024). While originally developed for oncology, similar frameworks are being extended to intracellular pathogens, where multi-pathway interference may be required to overcome robust adaptive mechanisms.

3.8.4 Metabolic disorders

Metabolic syndrome and type 2 diabetes mellitus represent additional domains where multi-target pharmacology is particularly relevant (Sarkar et al., 2023; Acharya et al., 2023; Roy et al., 2023; Biswas et al., 2023; Sur et al., 2023; Sadhu et al., 2025). These conditions encompass insulin resistance, dyslipidemia, low-grade inflammation, endothelial dysfunction, and often nonalcoholic fatty liver disease, all of which contribute to cardiovascular risk.

Multi-isoform PPAR modulators, designed to activate PPARα, PPARγ and PPARδ to varying degrees, aim to simultaneously improve lipid profiles, insulin sensitivity, and inflammatory status beyond what can be achieved with selective PPARγ agonists alone. RB394, a dual sEH/PPARγ modulator, exemplifies a rational multi-target agent. By inhibiting soluble epoxide hydrolase and activating PPARγ, RB394 reduces inflammation, improves insulin sensitivity, and limits weight gain through effects on ENaC channels and epoxyeicosatrienoic acid levels in rodent models of type 2 diabetes (Lillich et al., 2021). In ZSF1 rats, this compound ameliorates hyperglycemia, hyperlipidemia, and insulin resistance while also attenuating renal fibrosis and hepatic steatosis, indicating multi-organ protection.

Commonly used antidiabetic drugs often exhibit inherent polypharmacology. Metformin, in addition to its primary effects on hepatic gluconeogenesis via AMPK activation, also reduces plasminogen activator inhibitor-1 levels and confers cardiovascular benefits independent of glycemic control. Likewise, GLP-1-related mechanisms engaged by drugs like teneligliptin influence insulin secretion, glucagon release, gastric motility, appetite, and body weight, yielding a multilayered therapeutic profile (Lillich et al., 2021) (Figure 3.5).

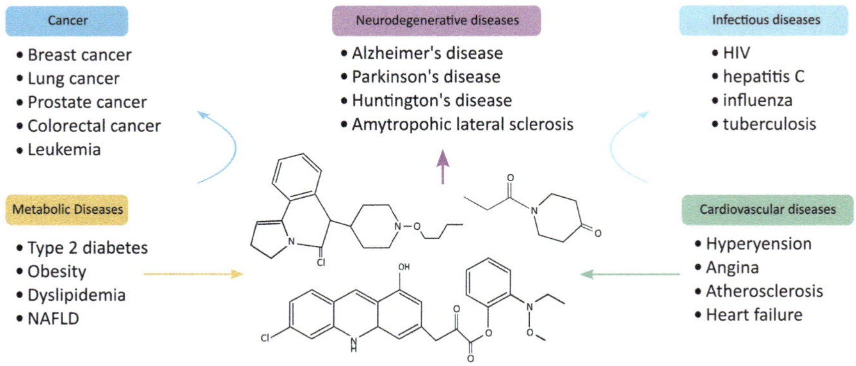

Figure 3.5 Multi-target drugs across disease areas.

3.9 CASE STUDIES OF SUCCESSFUL MULTI-TARGET DRUGS

Imatinib (Gleevec) was originally designed as a selective inhibitor of the BCR–ABL fusion kinase in CML but was subsequently found to inhibit PDGFR, c-Kit, and additional tyrosine kinases. This broader activity profile enabled extension of its clinical utility to GIST and other malignancies driven by KIT or PDGFR signaling. Initially regarded as potential off-target liabilities, these additional activities were ultimately reframed as therapeutically advantageous polypharmacology, illustrating how unrecognized multi-target profiles can create new indications when appropriately aligned with disease biology. Kinase panel profiling has confirmed that imatinib interacts with dozens of kinases at therapeutically relevant concentrations, highlighting that effective targeted agents often operate within a polypharmacological space rather than as perfectly selective inhibitors.

Aspirin and other NSAIDs provide another instructive case. While their classical mechanism involves inhibition of COX-1 and COX-2 enzymes and consequent suppression of prostaglandin and thromboxane biosynthesis, broader screening has revealed interactions with numerous kinases, modulation of NF-κB signaling, activation of PPAR-γ, and effects on various ion channels and GPCRs. These additional mechanisms likely contribute to aspirin's cardioprotective and chemopreventive properties, which cannot be fully accounted for by COX inhibition alone. Recognition of this multi-target profile has stimulated interest in developing NSAID-based derivatives that preserve beneficial anti-inflammatory, antithrombotic, and anticancer properties while minimizing gastrointestinal and renal adverse effects.

Thalidomide and its derivatives, lenalidomide and pomalidomide, illustrate how adverse multi-target effects can be repurposed into powerful therapeutic tools. Thalidomide's initial withdrawal due to teratogenicity

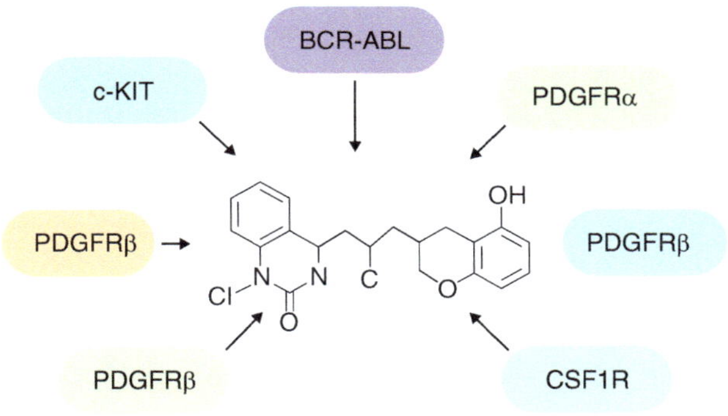

Figure 3.6 Polypharmacology of imatinib.

obscured its complex immunomodulatory actions. Subsequent work revealed that it binds cereblon (CRBN), a component of the CRL4-CRBN E3 ubiquitin ligase complex, leading to targeted degradation of transcription factors such as IKZF1 and IKZF3, alongside modulation of cytokine production and T cell co-stimulation. These multifaceted immunological effects underpin its efficacy in multiple myeloma and related malignancies. The evolving understanding of CRBN-mediated substrate degradation has transformed these agents into prototypes for proteolysis-targeting strategies.

Sorafenib provides an example of how extended polypharmacology can simultaneously drive efficacy and toxicity. Beyond its inhibition of VEGFR, PDGFR, Raf kinases, c-Kit, and Flt-3, sorafenib also affects unrelated targets, including p38 MAPK, SGK, and Aurora kinases (Sunder et al., 2023). While this broad activity contributes to its antitumor properties, it is also implicated in adverse events such as hypertension, hepatotoxicity, and cardiac dysfunction. Dissecting sorafenib's full target spectrum has facilitated efforts to design next-generation multi-target compounds that preserve critical anticancer activities while attenuating toxicity through improved selectivity and optimized binding modes (Figure 3.6).

3.10 CHALLENGES AND LIMITATIONS

Despite its promise, multi-target drug design is accompanied by formidable conceptual, practical, and translational challenges.

One major difficulty is target selection. Identifying optimal combinations of targets from the large number of druggable proteins is computationally and experimentally complex. Even restricting attention to 500 candidate targets, the number of possible pairs exceeds 124,000, and triplet combinations exceed 20 million, making exhaustive evaluation infeasible

(Tang et al., 2014). Moreover, incomplete knowledge of disease networks—particularly in rare or poorly characterized conditions—limits the reliability of predictive models.

A second challenge lies in medicinal chemistry optimization. Multi-target molecules must meet potency thresholds at all intended targets while simultaneously avoiding anti-targets that confer toxicity. Balancing these competing demands in high-dimensional property space—where potency, selectivity, solubility, permeability, metabolic stability, and synthetic tractability must all be maintained—is substantially more difficult than optimizing for a single target. Although machine learning and generative models offer powerful guidance, they do not eliminate the need for extensive experimental iteration.

ADMET properties impose further constraints. Achieving simultaneous, therapeutically relevant exposure at multiple targets often favors larger, more lipophilic compounds, but these properties can also increase risks of hepatotoxicity, hERG channel inhibition (and consequent QT prolongation), and promiscuous binding. The tension between broad multi-target engagement and acceptable safety profiles is therefore a recurring obstacle (Luo et al., 2013).

Polypharmacology inherently increases the probability of adverse events arising from unintended interactions. Multi-target kinase inhibitors, for instance, may inhibit essential kinases in non-malignant tissues, leading to cardiovascular, dermatologic, or gastrointestinal toxicities (Sunder et al., 2023). Disentangling which targets contribute to efficacy and which drive toxicity is challenging, but necessary for rational refinement.

Finally, regulatory and developmental pathways for multi-target drugs are still evolving. Current guidelines were largely developed around single-target agents and do not explicitly address evaluation of intentional multi-target mechanisms. Uncertainty about how regulators will weigh the trade-off between increased complexity and potential therapeutic gain can complicate development planning, trial design, and risk–benefit assessment.

3.11 REGULATORY AND SAFETY CONSIDERATIONS

Regulatory agencies such as the U.S. Food and Drug Administration (FDA) are increasingly cognizant of the polypharmacological nature of many modern therapeutics but have not yet fully codified frameworks specific to deliberate multi-target agents. Guidance documents for safety pharmacology, such as ICH S7A, emphasize thorough assessment of cardiovascular, central nervous system, and respiratory safety and require extensive preclinical profiling, particularly for small-molecule kinase inhibitors (U.S. Food and Drug Administration, 2001).

For kinase-directed therapies, comprehensive kinase panel screening is typically expected, often encompassing several hundred kinases, to help characterize selectivity profiles and anticipate potential off-target liabilities (Xiong et al., 2021). Particular attention is directed to cardiac safety, including hERG channel inhibition and QTc prolongation, as well as hepatotoxicity and other organ-specific toxicities.

In the context of multi-target drugs, regulators increasingly expect demonstration that simultaneous engagement of multiple targets provides clinically meaningful benefits that justify added complexity and possible risks. Adaptive clinical trial designs and biomarker-based patient stratification are being employed to identify subgroups that derive maximal advantage from multi-target therapies, while minimizing exposure in patients for whom the benefit–risk ratio is less favorable.

Post-marketing surveillance remains essential for detecting rare or long-latency toxicities associated with complex multi-target profiles. When unexpected adverse events emerge, mechanistic follow-up studies are often required to determine whether they stem from intended pharmacology or from unanticipated off-target actions. Guidance on reproductive and developmental toxicity (DART) further underscores the need to investigate biologically plausible risks even in the absence of an initial mechanistic hypothesis (U.S. Food and Drug Administration, 2021). For intentional polypharmacology, robust experimental evidence dissecting target-wise contributions to efficacy and toxicity—using genetic knockouts, selective modulators, and in vivo occupancy studies—is increasingly important to support risk–benefit assessments and labeling decisions (Table 3.4).

Table 3.4 Key regulatory expectations for multi-target drugs

Requirement	*Description*
Kinase panel screening	Broad profiling across extensive kinase panels to assess target selectivity and identify potential off-target interactions.
Cardiac safety	Evaluation of cardiotoxic risk using hERG channel inhibition assays, QT/QTc interval assessment, and related electrophysiological studies.
Biomarker-driven trials	Use of predictive and pharmacodynamic biomarkers to stratify patients and optimize the benefit–risk profile in clinical trials.
Post-marketing monitoring	Long-term pharmacovigilance to detect rare adverse events, cumulative toxicity, and unintended off-target effects after approval.

3.12 FUTURE PERSPECTIVES AND EMERGING TRENDS

The future of multi-target and polypharmacology-oriented drug discovery is likely to be shaped by deep integration of AI, multi-omics technologies, and precision medicine concepts.

AI-driven design is expected to further accelerate identification of multi-target compounds. Deep generative models, graph neural networks, and reinforcement learning algorithms can jointly explore chemical and target spaces, optimizing structures for desired multi-target profiles while penalizing predicted interactions with high-risk anti-targets (Munson et al., 2024; Cichońska et al., 2024). Continued improvements in model interpretability, incorporation of three-dimensional structural information, and integration of synthetic feasibility constraints will make AI-generated designs more practical and impactful.

Concurrently, systems-level biomarker discovery is enabling more refined mapping of disease networks. High-throughput proteomics and phosphoproteomics can delineate dysregulated signaling modules at unprecedented resolution, while integration of genomic, transcriptomic, and metabolomic data yields multilayered disease signatures (Hartl et al., 2021). These datasets inform the selection of target combinations and help define pharmacodynamic readouts that capture multi-target engagement in patients.

CRISPR-based genetic screens and synthetic lethality studies are becoming major tools for uncovering synergistic target pairs or networks. By systematically perturbing pairs or sets of genes in cancer cell models, researchers can identify combinations whose concurrent disruption produces cell death or growth arrest far exceeding that seen with individual perturbations (Xiong et al., 2021). Such data support rational multi-target drug design tailored to specific genomic contexts, offering a route toward truly precision polypharmacology.

Conceptual models in network pharmacology are also evolving. Rather than focusing solely on catalytic enzymes or canonical receptors, increasing attention is being directed to scaffold proteins, adaptors, and other structural nodes that mediate crosstalk between pathways. Targeting such hubs may enable efficient remodeling of disease networks with fewer direct interventions. In parallel, combination strategies that integrate multi-target small molecules with immunotherapy or cell-based therapies are likely to expand. For example, pairing pathway-targeted agents that enhance antigen presentation or reduce immunosuppressive signals with immune checkpoint inhibitors could yield synergistic improvements in antitumor immunity.

Overall, future therapeutic paradigms are expected to shift from treating single pathways in isolation to modulating disease networks in a coordinated and context-dependent manner.

3.13 CONCLUSION

Multi-target and polypharmacology-oriented drug design represents a significant departure from traditional single-target discovery strategies and reflects a broader shift toward systems-level thinking in biomedical research. The complex, adaptive, and multifactorial nature of most major human diseases has highlighted the limitations of monotherapy aimed at individual targets and has underscored the need for therapeutics capable of orchestrated network modulation.

By intentionally engaging multiple nodes within disease-associated networks, multi-target drugs can address pathway redundancy, mitigate compensatory bypass mechanisms, and reduce the likelihood of resistance. At the same time, they offer practical advantages in terms of simplified dosing, improved adherence, and unified pharmacokinetic profiles compared with multi-drug regimens.

Successful implementation of this paradigm requires tight integration of network pharmacology, computational modeling, medicinal chemistry, and rigorous experimental validation. Recent advances in AI, multi-omics profiling, and functional genomics are expanding the toolkit available for rational design and prioritization of multi-target strategies. Nonetheless, substantial challenges remain, including target selection under uncertainty, chemical optimization in high-dimensional property space, and careful management of safety and regulatory complexities.

As these challenges are progressively addressed, multi-target approaches are poised to become increasingly central in the treatment of complex diseases. Single-target agents will likely retain their importance for certain monogenic or narrowly defined indications, but the therapeutic landscape will increasingly favor interventions that recognize and exploit the interconnected architecture of biological systems, aiming for more durable and comprehensive clinical benefit.

REFERENCES

Acharya, C.K., Das, B., Madhu, N.R., Sau, S., Manna De, M., & Sarkar, B. (2023). *A Comprehensive Pharmacological Appraisal of Indian Traditional Medicinal Plants with Anti-diabetic Potential*. Advances in Diabetes Research and Management. Springer, Singapore. pp. 163–193. https://doi.org/10.1007/978-981-19-0027-3_8

Agamah, F.E., Mazandu, G.K., & Mulder, N. (2019). Computational/in silico methods in drug target and lead compound identification. *Brief. Bioinformatics 20*(3), 854–871.

Anighoro, A., Bajorath, J., & Rastelli, G. (2014). Polypharmacology: challenges and opportunities in drug discovery. *J. Med. Chem., 57*(19), 7874–7887. https://doi.org/10.1021/jm5006463

Biswas, G., Madhu, N.R., Sarkar, B., Paul, S., Erfani, H., & Alam, Q. (2024). Rare Genetic Disorders: Unraveling the Pathophysiology, Gene Mutations, and Therapeutic Advances in Fabry Disease and Marfan Syndrome. In: Umair, M., Rafeeq, M., Alam, Q. (eds) *Rare Genetic Disorders: Advancements in Diagnosis and Treatment*. Springer, Singapore. pp. 199–219. https://doi.org/10.1007/978-981-99-9323-9_7

Biswas, T., Behera, B.K., & Madhu, N.R. (2023). *Technology in the Management of Type* 1 *and Type* 2 *Diabetes Mellitus: Recent Status and Future Prospects.* Advances in Diabetes Research and Management. Springer, Singapore. pp. 111–136. https://doi.org/10.1007/978-981-19-0027-3_6

Boran, A.D.W., Dolph, M.J., & Xia, B. (2010). Systems approaches to polypharmacology and drug discovery. *Wiley Interdiscipl. Rev. Syst. Biol. Med., 2*(2), 194–209.

Cheng, F., Kovacs, I.A., & Barabasi, A.L. (2019). Network-based prediction of drug combinations. *Nat. Commun., 10*, 1197. https://doi.org/10.1038/s41467-019-09186-x

Chopade, P., Prabhudesai, S., Prabhudesai, H., & Hussain, I. (2022). Alzheimer's and Parkinson's disease therapies in the clinic and in the pipeline. *Front. Mol. Biosci., 9*, 925841.

Cichońska, A., Ravikumar, B., Allara, M., & Tomlinson, W. (2024). AI for targeted polypharmacology: the next frontier in drug discovery. *Curr. Opin. Struct. Biol., 82*, 102631. https://doi.org/10.1016/j.sbi.2023.102771

Das, S.K., Sen, K., Sanyal, T., Saha, A., & Madhu, N.R. (2024). Flavonoids: A Promising Neuroprotectant and Its Salutary Effects on Age-Related Neurodegenerative Disorders. In: Pathak, S., Banerjee, A. (eds) *Neuroprotective Effects of Phytochemicals in Brain Ageing*. Springer Nature Singapore: Singapore. pp. 221–255. https://doi.org/10.1007/978-981-99-7269-2_11

Eron, J.J. (2000). HIV-1 protease inhibitors. *Clin. Infect. Dis., 30*(Supplement 2), S160–S170. https://doi.org/10.1086/313853

Gharbi, Y., & Mercado, R. (2024). A comprehensive review of emerging approaches in machine learning for de novo PROTAC design. *Digital Discovery, 3*(11), 2158–2176. https://doi.org/10.1039/D4DD00177J

Gridelli, C., Maione, P., & Rossi, A. (2007). Sorafenib and sunitinib in the treatment of advanced non-small-cell lung cancer. *Clin. Lung Cancer 8*(4), 235–242. https://doi.org/10.3816/CLC.2007.n.023

Hartl, D., Eckert, R., & Reymond, L. (2021). Translational precision medicine: an industry perspective. *Curr. Opin. Chem. Biol., 61*, 140–147.

He, X., & Xu, C. (2020). Immune checkpoint signaling and cancer immunotherapy. *Cell Res. 30*(8), 660–669. https://doi.org/10.1038/s41422-020-0343-4

Hopkins, A.L. (2008). Network pharmacology: the next paradigm in drug discovery. *Nat. Chem. Biol., 4*(11), 682–690. https://doi.org/10.1038/nchembio.118

Hsin, K.Y., Ghosh, S., & Kitano, H. (2013). Combining machine learning systems and multiple docking simulation packages to improve docking prediction reliability for network pharmacology. *PLoS ONE, 8*(12), e83922. https://doi.org/10.1371/journal.pone.0083922

Iida, M., Kawano, R., Igashira, A., & Tanaka, S. (2024). A network-based trans-omics approach for predicting drug combinations. *Commun. Biol., 7*, 571. https://doi.org/10.1038/s43856-024-00571-2

Kahraman, M., & Ozturk, S. (2019). Experimental study of newly structural design grinding wheel considering response surface optimization and Monte Carlo simulation. *Measurement, 147*, 106825. https://doi.org/10.1016/j.measurement.2019.07.053

Kant, S., Deepika, & Roy, S. (2025). Artificial intelligence in drug discovery and development: transforming challenges into opportunities. *Discover Pharmaceutical Sci., 1*, Article 7. https://doi.org/10.1007/s44395-025-00007-3

Kim, A.R., Litzenberg, D., & Griesinger, F. (2009). Sorafenib and sunitinib: multikinase inhibitors for cancer. *Oncol. Pharm. Pract., 15*(1), 47–62.

Lei, P., Ayton, S., & Bush, A.I. (2011). GSK-3 in neurodegenerative diseases. *Int. J. Alzheimers Dis., 2011*, 189246. https://doi.org/10.4061/2011/189246

Lillich, F.F., Demir, O., & Lah, M.S. (2021). Multi-target approaches in metabolic syndrome and type 2 diabetes. *Front. Pharmacol., 12*, 624505. https://doi.org/10.3389/fphar.2020.554961

Luo, Y., Madden, M., & Sun, Y. (2013). Contributions of molecular properties to drug promiscuity. *J. Med. Chem., 56*(6), 2231–2246.

Madhu, N.R., Biswas, G., Paul, S., Adhikari, S., Sarkar, B., Rafeeq, M.M., & Umair, M. (2024). Challenges and Future Opportunities in Rare Genetic Disorders: A Comprehensive Review. In: Umair, M., Rafeeq, M., Alam, Q. (eds) *Rare Genetic Disorders*. Springer, Singapore. pp. 251–275. https://doi.org/10.1007/978-981-99-9323-9_9

Madhu, N.R., Sarkar, B., Biswas, P., Roychoudhury, S., Behera, B.K., & Acharya, C.K. (2024). Therapeutic Potential of Melatonin in Glioblastoma: Current Knowledge and Future Prospects. In: Sobti, R.C., Watanabe, M., Sobti, A. (eds), *Biomarkers in Cancer Detection and Monitoring of Therapeutics: Volume 2, diagnostic and therapeutic applications* (pp. 371–386). Academic Press. https://doi.org/10.1016/B978-0-323-95114-2.00002-9

Madhu, N.R., Sarkar, B., Halder, N.C., & Behera, B.K. (2023). Potential Applications of Traditional Medicinal Plants for Treating Sleep Disorders. In: Sarkar, B., Biswas, S.J., Samal, A.C., Pandey, A. (eds) *The Basic Handbook of Indian Ethnobotany and Traditional Medicine* [Volume: 2]. International Academic Publishing House. pp. 224–243. https://doi.org/10.52756/bhietm.2023.e02.015

Madhu, N.R., Sarkar, B., Roychoudhury, S., Behera, B.K. (2023). Melatonin Induced in Cancer as a Frame of Zebrafish Model. In: Pathak, S., Banerjee, A., Bisgin, A. (eds) *Handbook of Animal Models and Its Uses in Cancer Research*. Springer, Singapore. pp. 1–18. https://doi.org/10.1007/978-981-19-3824-5_61

Maenza, J., Flexner, C., & Sperber, S. (1998). Combination antiretroviral therapy for HIV infection. *Am. Fam. Physician, 57*(11), 2789–2800.

Morphy, R., Kay, C., & Rankovic, Z. (2006). From magic bullets to designed multiple ligands. *Drug Discov. Today, 11*(13–14), 641–648.

Munson, B.P., Karunasena, H.C., & Harki, D.A. (2024). De novo generation of multi-target compounds using deep generative and graph neural networks. *Nat. Commun., 15*, 4182. https://doi.org/10.1038/s41467-024-47120-y

Paoli, C.J., McIntosh, A., & Williams, D. (2024). Effectiveness of single-tablet combination therapy in improving adherence and persistence and their economic outcomes. *J. Health Econ. Outcomes Res. 11*(1), 23–48. https://doi.org/10.36469/001c.91396

Ramirez, M., Rajaram, S., Steininger, R.J., Osipchuk, D., & Roth, M.A. (2016). Diverse drug-resistance mechanisms can emerge from drug-tolerant cancer persister cells. *Nat. Commun.*, *7*, 10690. https://doi.org/10.1038/ncomms10690

Reddy, A.S., Zhang, S., & Ponatowski, R. (2013). Polypharmacology: drug discovery for the future. *Expert Rev. Clin. Pharmacol.*, *6*(1), 41–47. https://doi.org/10.1586/ecp.12.74

Roy, P., Ghosh, D., Sanyal, R., Madhu, N.R., & Dey, A. (2024). The Controversy Surrounding Drugs against Neurodegenerative Disorders: Benefit or Harm? In: Pathak, S., Banerjee, A. (eds) *Neuroprotective Effects of Phytochemicals in Brain Ageing*. Springer, Singapore. pp. 373–386. https://doi.org/10.1007/978-981-99-7269-2_17

Roy, R., Chakraborty, A., Jana, K., Sarkar, B., Biswas, P., & Madhu, N.R. (2023). *The Broader Aspects of Treating Diabetes with the Application of Nanobiotechnology*. Advances in Diabetes Research and Management. Springer, Singapore. pp. 137–162. https://doi.org/10.1007/978-981-19-0027-3_7

Ryszkiewicz, P., Wójcik, M., & Szczepanik, M. (2023). Polypharmacology: promises and new drugs in 2022. *Int. J. Mol. Sci.*, *24*(4), 3873. https://doi.org/10.1007/s43440-023-00501-4

Sadhu, S., Sarkar, B., Paul, T., Sanyal, T., Mitra, A., & Madhu, N.R. (2025). The potential of stem cells in the treatment of diabetes: an up-to-date review. *Biomed. Res. Ther.*, *12*(8), 7680–7696. https://doi.org/10.15419/m7mc9595.

Sarkar, S., Sadhu, S., Roy, R., Tarafdar, S., Mukherjee, N., Sil, M., Goswami, A., & Madhu, N.R. (2023). *Contemporary drifts in diabetes management. Int. J. App. Pharm.*, *15*(2), 1–9. https://doi.org/10.22159/ijap.2023v15i2.46792

Shepherd, C., Karamitsos, D., & Rahman, S. (2022). Surface plasmon resonance screening to identify active and selective kinase inhibitors. *J. Biomol. Screen.*, *27*(2), 121–134.

Sunder, S.S., Ewer, M.S., & Vaidya, P. (2023). Adverse effects of tyrosine kinase inhibitors in cancer therapy. *Signal Transduct. Target. Ther. 8*, 15. https://doi.org/10.1038/s41392-023-01469-6

Sur, T., Das, A., Bashar, S., Tarafdar, S., Sarkar, B., & Madhu, N.R. (2023). *Biochemical Assay for Measuring Diabetes Mellitus*. Advances in Diabetes Research and Management. Springer, Singapore. pp. 1–20. https://doi.org/10.1007/978-981-19-0027-3_1

Sydow, D., Burggraaff, L., & Szengel, A. (2019). Advances and challenges in computational target prediction. *J. Chem. Inf. Model.*, *59*(5), 1728–1742. https://doi.org/10.1021/acs.jcim.8b00832

Tang, J., Aittokallio, T., & Ravikumar, B. (2014). Network pharmacology strategies toward multi-target anticancer drug discovery. *Genome Biol. 15*(6), 417.

U.S. Food and Drug Administration. (2001). *Guidance for industry: S7A safety pharmacology studies for human pharmaceuticals*. U.S. Department of Health and Human Services.

U.S. Food and Drug Administration. (2021). *S5(R3) detection of reproductive and developmental toxicity for human pharmaceuticals: Guidance for industry*. U.S. Department of Health and Human Services.

Van der Schyf, C.J., Geldenhuys, W.J., & Youdim, M.B. (2011). The use of multi-target drugs in the treatment of neurodegenerative diseases. *Expert Rev. Clin. Pharmacol.*, *4*(2), 293–298. https://doi.org/10.1586/ecp.11.13

Wojtukiewicz, M.Z., Sierko, E., & Grela-Wojewoda, A. (2021). Inhibitors of immune checkpoints-PD-1, PD-L1, CTLA-4-new opportunities for cancer patients. *Cancers, 12*(5), 1163. https://doi.org/10.1007/s10555-021-09976-0

Xiong, Z., Volegova, M., & Lyons, K. (2021). Crowdsourced identification of multi-target kinase inhibitors for RET-dependent tumors and Tauopathy. *PLoS Comput. Biol., 17*(9), e1009325. https://doi.org/10.1371/journal.pcbi.1009302

Zhao, S., & Iyengar, R. (2012). Systems pharmacology: network analysis to identify multiscale mechanisms of drug action. *Annu. Rev. Pharmacol. Toxicol., 52*, 505–521. https://doi.org/10.1146/annurev-pharmtox-010611-134520

Chapter 4

Computational strategies for cancer drug discovery

Sajana Mouvanal, Moganamurthy Brunda, Mutthuraj Dasegowda, Kaustubha Beliya, K. M. Varun, M. Vasundhara, and Kanthesh M. Basalingappa

Division of Molecular Biology, School of Life Sciences, JSS Academy of Higher Education & Research, Mysuru, Karnataka

4.1 INTRODUCTION

Cancer is one of the most complicated diseases, which is fueled by genetic mutations, malfunctioning signaling, metabolic restructuring, and interactions within the tumor microenvironment [1]. With over 90% of oncology drug discovery programs, on average, failing to reach the clinic, conventional drug discovery techniques such as target identification, lead optimization, preclinical testing, and clinical validation are not only costly but also time-consuming [2].

In order to break these bottlenecks, the drug discovery paradigm has changed to employ computational strategies [2]. Components integrated in the discovery process, such as cheminformatics, structural biology, systems biology, and artificial intelligence (AI), facilitate rational and data-driven decision-making throughout discovery [3]. The computational tools are helpful in predicting drug-target interactions (DTIs), pharmacokinetic and toxicity profiles, chemical scaffold optimization, and molecular dynamics (MD) simulation prior to experimental validation [4].

In the recent past, especially in machine learning (ML), deep learning (DL), and graph neural networks (GNNs), predictive modeling of drug efficacy and safety has been improved rapidly [5]. Molecular simulations, virtual screening (VS), and network pharmacology can be used to search the ample chemical space to find multi-target and mechanism-specific cancer medicines [3].

Moreover, there have been fresh considerations of the natural products-based leads due to their structural diversity and biological significance. Recent studies examining phytochemical interventions in leukemia and cancer stem cell pathways are especially indicative of this growing interest. The way bioactive compounds influence apoptosis, oxidative stress, and signaling cascades like PI3K/Akt and MAPK, for example, is highlighted in our recent review on acute myeloid leukemia (AML) [6]. Furthermore, new

 DOI: 10.1201/9781003672760-4

therapeutic insights into natural products that target stemness, differentiation, and drug resistance are covered in our book chapter on bioactive compounds and cancer stem cells. Combining high-resolution mass spectrometry (HR-LC-MS) information, molecular docking, and network-based screening computational workflows has become extremely useful in discovering novel bioactive phytochemicals.

Altogether, computational oncology is becoming a hybrid field between molecular pharmacology, bioinformatics, and AI. The chapter also describes significant computational models, such as structure-based and ligand-based design, MD, and VS, explaining how the models complement each another in the process of anticancer drug discovery (Figure 4.1).

4.1.1 Historical perspective and evolution of computational oncology

Computer-assisted drug design (CADD) is a conceptual idea that was developed in the 1980s when molecular docking and quantitative structure-activity relationship (QSAR) models were introduced [7]. Originally, these techniques were used to screen and optimize hits on a small scale. Progress in structural genomics, computational capability, and AI over the last 10 years has extended their harvests to the field of precise oncology, in which the treatment plan is determined concerning the patient-specific tumor transcriptome [8].

The most striking developments are the development of molecular docking engines (AutoDock and Glide) and MD software (AMBER and GROMACS), as well as the development of new AI-based predictors (AlphaFold2), which have revolutionized the prediction of protein structure [6]. The computational methods have saved time and money by directing and improving experimental drug discovery (Figure 4.2).

4.2 STRUCTURE-BASED DRUG DESIGN (SBDD)

Structure-based drug design (SBDD) uses the three-dimensional (3D) structure of a biological target to create ligands that bind with high affinity and specificity rationally. SBDD elucidates molecular recognition by simulating molecular recognition and enables optimization of candidate molecules on an atomic level [9].

4.2.1 Protein structure determination and modeling

In phylogenetic studies, determining and modeling protein structure play a crucial role. Protein structure determination and computational modeling play important roles in phylogenetic research by enabling comparisons of

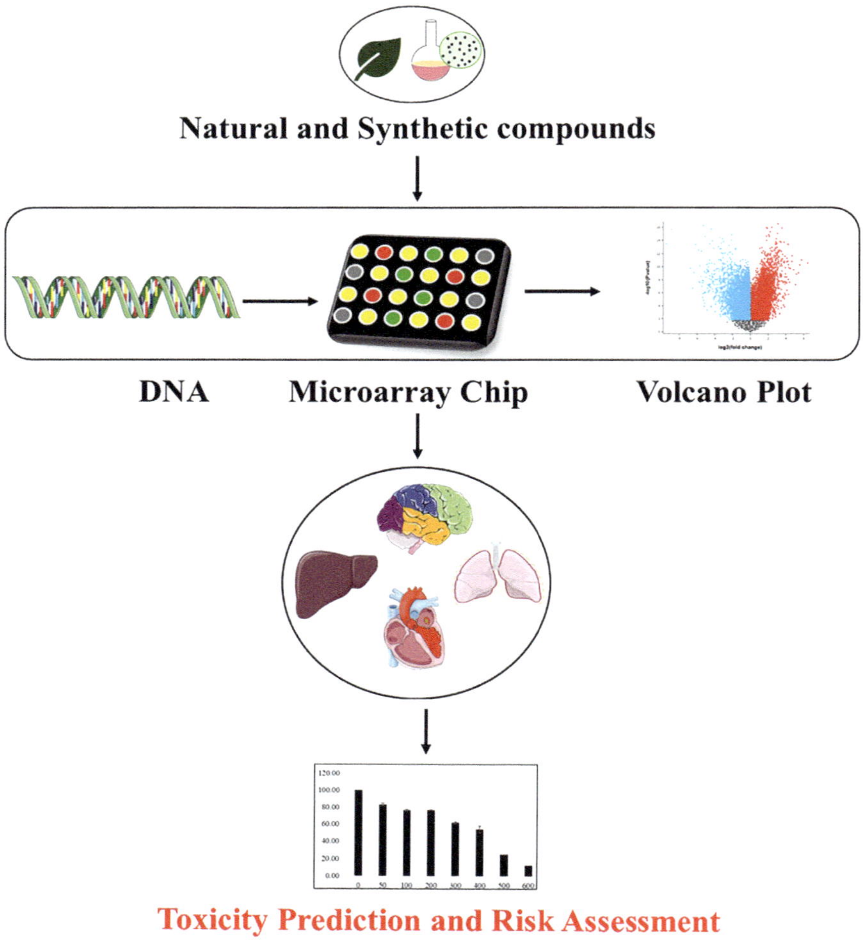

Figure 4.1 Workflow for toxicity prediction and risk assessment of natural and synthetic compounds.

structural conservation, evolutionary relationships, and functional divergence among proteins.

SBDD starts with proper knowledge of the 3D structure of the target. High-resolution structural data deposited in the Protein Data Bank (PDB) is obtained through experimental techniques such as X-ray crystallography, nuclear magnetic resonance (NMR) spectroscopy, and cryo-electron microscopy (cryo-EM) [10].

Nevertheless, experimental structures of not all cancer-related proteins have been solved. Computational modeling techniques, such as homology modeling (e.g., MODELLER, SWISS-MODEL) and ab initio modeling, are

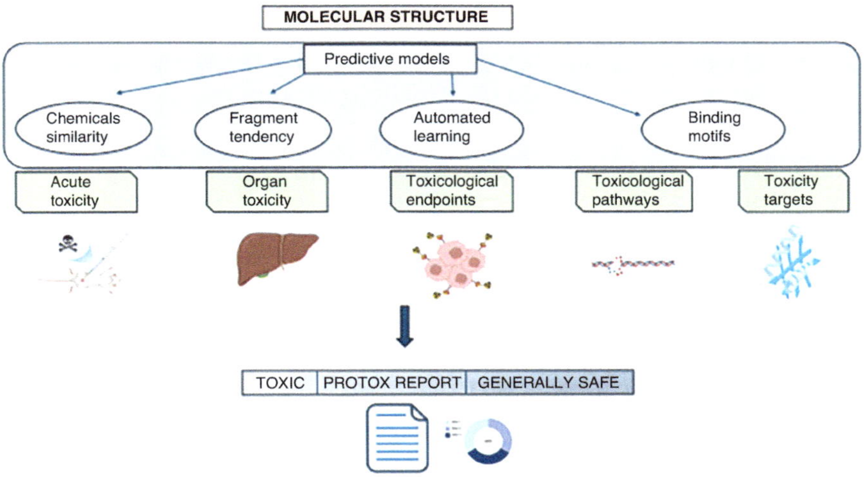

Figure 4.2 Workflow for computational toxicity prediction. Predictive models derived from molecular structure evaluate chemical similarity, fragment patterns, machine-learning features, and binding motifs to estimate acute toxicity, organ toxicity, toxicological endpoints, pathways, and molecular targets, resulting in a ProTox-style toxicity classification.

used to fill this gap. Most recently, AlphaFold2 and RoseTTAFold have both achieved a significant breakthrough in the accuracy of predicted structures, now making SBDD accessible to previously inaccessible targets [11].

After a protein model has been validated, ligand-accessible cavities can be defined using binding site identification tools like CASTp, SiteMap, or Fpocket. Subsequently molecular docking and VS processes can be implemented.

4.2.2 Molecular docking

Molecular docking is used to predict the preferred orientation of a ligand in a protein's active site and its binding affinity. The most widely used docking programs are AutoDock Vina, GOLD, Glide, and DOCK [12].

An ordinary docking process covers:

- **Protein and Ligand Preparation**: optimization, addition of hydrogens, and generating grids.
- **Pose Sampling**: ligand conformation and orientation sampling.
- **Scoring**: scoring of poses on the basis of empirical or machine-learned scoring functions.
- **Refinement after Docking**: rescoring with superior free energy methods or shorter simulations of the MD [13].

Docking is an essential component of VS programs, but its disadvantages, including unbending receptor approximations and the inability to consider solvent dynamics, require that it be paired with other methods, like MD simulations, to bring increased realism to the simulation.

4.2.3 Hotspots and site characterization: binding sites

The characterization of binding sites is important in the study of protein-ligand interaction energetics. Hotspots in the dynamics of computational mapping (e.g., FTMap, SiteHound) are used to locate the hottest spots in the residues in binding energy. These hotspots inform medical chemists how to rationally modify scafolds, use fragment-based design, and optimize by combinatorics [14].

Hotspot analyses have also played a crucial role in cancer biology in the identification of important residues in protein kinases, apoptosis controllers (e.g., Bcl-2 family), and DNA repair factors (e.g., CHK1, ATR) [15]. This knowledge is the foundation of the design of targeted therapy and the explanation of resistance mechanisms.

4.3 DESIGNING A LIGAND-BASED DRUG (LBDD)

If the 3D structure of the target is not available or reliable, ligand-based drug design (LBDD) can be used as an alternative. Finding patterns connected to the activity of known active compounds and creating new, possibly more potent analogs relies on examining their structural and physicochemical properties [16].

4.3.1 Pharmacophore modeling

A pharmacophore is the geometrical arrangement of biological activity-related features like hydrogen-bond acceptors/donors, hydrophobic centers, and aromatic rings. Pharmacophore modeling is conducted with the help of tools such as LigandScout, MOE, and PharmaGist [17].

Ligand-based pharmacophore models are based on active compounds, whereas structure-based models are obtained based on protein-ligand complexes. Pharmacophore-based VS enables the efficient screening of new hits in a large subset of compounds [18].

4.3.2 Quantitative structure-activity relationship (QSAR) model

QSAR modeling is a technique that relates molecular descriptors to biological activity using either statistical or ML techniques [19]. The description sets can be topological indices, physicochemical values, or molecular fingerprints (e.g., ECFP4).

The ML methods applied in modern QSAR include Deep Neural Networks, Gradient Boosting, Support Vector Machines, and Random Forests. The models include predicting activities (IC50, Ki) or classification results (active/inactive). The cross-validation and Y-randomization of the model are to validate its robustness. The hybrid methods, which involve the use of QSAR along with docking and pharmacophore models, enhance predictive accuracy [20].

LBDD has been especially effective in natural product studies, in which structural diversity of phytochemicals allows scaffold hopping and bioactivity-directed optimization [21].

4.4 MOLECULAR DYNAMICS (MD) SIMULATIONS AND FREE-ENERGY METHODS

The MD simulations provide an extension between the outcomes of the static docking and the dynamics of the biomolecule behavior by treating atomic motions and conformational freedom in a physiological environment [22].

4.4.1 Classical MD simulations

Classical MD relies on Newtonian mechanics to model the molecular system with the aid of force fields such as AMBER, CHARMM, and OPLS. The operating principle of the algorithm includes the step of system solvation, ionization, energy optimization, equilibration, and production cycles (typically 50–500 ns) [23].

Trajectory analysis provides data on structural stability (RMSD, RMSF), compactness (radius gyration), hydrogen bonding, and the breakdown of interaction energy. In cancer studies, MD has been widely and regularly used to optimize docking outcomes, determine binding stability, and allow induced-fit mechanisms in enzyme or receptor targets to be observed [24].

4.4.2 Improved sampling and free-energy calculations

Traditional MD can only be affected by reachable timescales, whereby improved sampling methods are utilized: metadynamics, accelerated MD, and replica exchange MD (REMD) are used to examine unfavorable events and conformational transitions [25].

Calculations of free energy, such as MM/PBSA and/or MM/GBSA and Free-Energy Perturbation (FEP), to determine ligand binding affinities can be estimated more precisely and can quantitatively rank out hits [26]. This is especially useful in the discovery of selective inhibitors of kinases, proteases, and enzymes that bind DNA that are associated with cancer development.

In order to model large biomolecular assemblies or long-timescale effects, coarse-grained (CG) models, simplify the atomistic description of a biomolecule while maintaining key dynamics. Hybrid quantum-classical (QM/MM) MD simulations provide atomistic resolution at catalytic or binding locations and are computationally effective [27]. The techniques prove helpful in the modeling of enzyme-ligand interactions as well as the mechanism of covalent inhibition about anticancer targets.

4.5 VIRTUAL SCREENING VIRTUAL EXPLORATION OF CHEMICAL SPACE

VS is a method that combines both structure- and ligand-based screening to filter large chemical libraries and find possible bioactive molecules [28]. VS has emerged as an important instrument in drug discovery for synthetic and natural products.

4.5.1 High-throughput virtual screening (HTVS)

High-throughput VS facilitates computational assessment of the many millions of compounds with parallelized docking processes on both cloud and GPU systems [29]. The most common types of screening pipelines are combinations of:

Pre-Filtering: the use of drug-likeness (Lipinski rules), lead-likeness, and PAINS filters.

Docking: the technique used involves CG rapid scoring, after which it is accurately docked.

Rescoring: with ML, free energy-based re-ranking.

Hit Clustering and Selection: grounded on the binding modes and scaffold diversity [30].

A combination of HTVS and absorption, distribution, metabolism, excretion, and toxicity (ADMET) prediction software (e.g., pkCSM, SwissADME) facilitates prioritization of early-stage candidates.

4.5.2 Virtual screening based on natural products

The importance of natural products as anticancer scaffolds is unquestionable because of their structural variety and biological applicability. This approach is in line with the integrative framework provided in our most recent review with an AML focus, wherein network pharmacology and molecular docking were integrated to rationalize the mechanism of action of phytochemicals in leukemia models [6].

Identification of promising phytochemicals can be done through computational screening of natural product databases (e.g., COCONUT, NPASS, SuperNatural II) [31].

Phytochemical VS usually involves the use of a combination of metabolomic annotation (LC-MS/MS, GNPS), docking, and pharmacophore filtering to predict bioactivity. Network pharmacology is an extension, demonstrating any possible multi-target effects of natural compounds [32].

Integration of Virtual Screening and Experimental Validation. This section indicates that the VS will be combined with experimental validation.

Computational hits are then tested in an experimental setting (to guarantee translational relevance) on in vitro assays like cytotoxicity assays in a cell line of leukemia or carcinoma cells. This in silico-in vitro loop is useful in increasing the reliability of computational predictions and gives way to hypothesis-driven discovery [33]. A combined HR-LC-MS profiling strategy and docking and MD refining is a powerful contemporary strategy of phytochemical-based anticancer lead discovery [34].

4.6 NETWORK PHARMACOLOGY AND MULTI-TARGET DESIGN

Cancer is a complex disease of systems and does not constitute a single-gene disorder but is interrelated in its signaling networks, redundancy, and compensation systems. Conventional one-drug-one-target paradigms do not have the capacity to address the heterogeneity of cancer. Systems biology network pharmacology combines systems biology, cheminformatics, and computational modeling to design or discover drugs that can simultaneously engage multiple nodes or pathways [35]. The results presented in our chapter on bioactive compounds and cancer stem cells, which focused on how multi-target phytochemicals interfere with stemness and oncogenic signaling through interdependent molecular networks, are in line with such systems-level approaches (Mouvanal & Basalingappa, 2024b).

4.6.1 Networks drug-target-disease augmentation

Network pharmacology is initially initiated by the assembly of drug-target-disease interaction maps. These networks combine multi-omics data (transcriptomics, proteomics, metabolomics) with data mining databases, including STRING, STITCH, DisGeNET, and CTD [36].

Workflow steps include the following:

Target Identification: differentially expressed genes/ proteins are mined out of cancer datasets.
Target-Disease Association: database mapping (OMIM, KEGG, Reactome).
Drug-Target Prediction: by docking, ligand-based similarity, and text mining.
Visualization and Analysis of Network: make the hub and cluster using Cytoscape or NetworkAnalyst.

Network types, such as degree, betweenness, and eigenvector centrality, are used to rank targets as hubs that have control in various oncogenic processes [37].

4.6.2 Polypharmacology and drug repurposing

Polypharmacology aims to create or discover compounds that interact with multiple targets of interest, enhancing efficacy and reducing resistance [38]. Multi-target ligands can be generated using computational techniques, including multi-objective optimization, graph-based DL, and scaffold hopping.

The models of ML are used to predict multi-target binding profiles by combining molecular fingerprints with protein feature embeddings. For example, AI-based kinase cross-reactivity prediction has allowed finding dual CHK1/ATR inhibitors and Bcl-2 family regulators faster [39].

Network-guided repurposing has been an effective synergy between polypharmacology and translational bioinformatics, in which approved drugs are re-examined to have other applications in cancer networks.

4.6.3 Systems biology in network pharmacology

Systems biology is a sub-specialization in phylogenetics, molecular biology, and evolutionary biology, as well as a branch of network biology in pharmacology.

Network pharmacology lacks the mechanistic depth that systems biology offers. It represents biological systems through dynamic and modular networks that combine signaling, metabolism, and gene regulation [40].

Using a systems-level simulation, emergent interaction patterns and possible intervention cross-validation points may be identified by overlaying computationally predicted DTIs on signaling networks (e.g., PI3K/Akt/mTOR, MAPK, DDR) [41]. The approaches lead to rational combination therapy design and biomarker-driven treatment approaches.

4.7 MACHINE LEARNING, DEEP LEARNING, AND AI IN DRUG DISCOVERY

Modern computational drug discovery is redefined by ML and AI. They allow humanistic pattern recognition, prediction of molecular properties, and de novo design using high-scale datasets [42].

Drug-Target Interaction (DTI) Prediction \n Predictive models for DTI utilize mathematical and machine learning algorithms to forecast or estimate binding effects without the immediate need for experimental data.

ML models acquire quantitative correlations between biological activities and molecular structures. Two prevailing paradigms are feature-based (using fingerprints or descriptors) and GNNs (using molecular graphs) [43].

DTI prediction based on AI leads to the discovery of new correlations between compounds and proteins, facilitating the process of identifying hits and evaluating drug risks. The model evaluation has also been standardized in oncology-specific DTI prediction using large-scale benchmarks such as CandidateDrug4Cancer [44].

Embedding based on sequence and graph-based molecular features has shown a great deal of improvement in hybrid architectures focused on predicting binding affinity and modeling selectivity [45].

4.7.1 De novo molecule generation generative modeling

New chemical compounds with the desired bioactivity profile are generated by generative AI models. Applications of variational autoencoders (VAEs), generative adversarial networks (GANs), reinforcement learning (RL), and transformer-based sequence models are now being used in cancer drug discovery [46].

Reward functions allow these models to optimize molecular properties, such as docking score, ADMET parameters, and synthetic accessibility. The de novo generation has resulted in selective kinases, DNA intercalators, and epigenetic modulators that are more drug-like [47].

4.7.2 AI in cancer drug response prediction and precision oncology

Precision oncology. In precision oncology, AI forecasts treatment reactions based on multi-omics integration (genomic, transcriptomic, proteomic information). DL models are associated with molecular properties of tumors and drug sensitivity (IC50 values) or resistance phenotype [48].

BI or AI-analyzed biomarker discovery determines sets of gene expression that are associated with response or toxicity. Most recently, hybrid quantum neural networks have demonstrated higher predictive accuracy in drug response prediction in heterogeneous cancer data [49].

The end objective is to facilitate adaptive therapy; the models will continuously improve themselves, and based on clinical data, they will suggest custom treatment regimens.

4.7.3 Federated and privacy-aware multicenter cancer AI

Federated learning enables institutions to indirectly train a distributed model without exchanging raw data to overcome the drawbacks of data privacy and heterogeneity [50]. The method increases generalization while preserving patient confidentiality.

Privacy-preserving algorithms, in conjunction with federated AI networks, are becoming a prime facilitator.

4.8 IN SILICO ADMET AND TOXICITY PREDICTION

Poor ADMET characteristics are still one of the primary causes of drug failure in late stages. Computational screening can be used at an early stage to screen out poor candidates before experimental testing [51].

Tools: The SwissADME, pkCSM, ADMETlab, and ProTox-II are all common examples of tools based on ML algorithms to predict bioavailability, solubility, toxicity, and cardiotoxicity [52].

4.8.1 Toxicogenomics and multi-organ toxicity simulations

Toxicogenomics combines information on genomes to predict chemical effects on genes. The AI models of toxicogenomics associate structural alerts with organ-specific toxicities such as hepatotoxicity, nephrotoxicity, and cardiotoxicity [53].

In cancer therapy, these techniques are used to differentiate on-target cytotoxic effects and systemic toxicities to enhance the estimation of the therapeutic index. Multi-organ simulation systems have become more integrated between PBPK (physiologically based pharmacokinetic) and ML to conduct a complete assessment of safety [54].

4.9 REPURPOSING DRUG REPURPOSING TRANSLATIONAL RELEVANCE

Drug repurposing (repositioning) involves attempts to find a new therapeutic application for an already known drug, using established safety and pharmacokinetic properties to reduce the development time [55].

4.9.1 Signature matching and transcriptomics approaches

The signature-based repurposing involves a comparison of the gene expression patterns in disease states with drug-induced patterns. Tools such as Connectivity Map (CMap) and LINCS L1000 detect drugs whose expression patterns revert those of the cancer phenotypes [56].

Network integration can improve predictive accuracy by projecting drug-induced transcriptional changes onto signaling and metabolic pathways of interest in tumor biology [57].

4.9.2 AI-driven repurposing

AI and ML methods facilitate the creation of multiple modes of data integration, which include chemical structure, omics, literature, and clinical data to forecast new drug-disease interactions [58].

Recent neural network-based models, including TxGNN, achieve this by engaging in zero-shot drug repurposing through learning transferable disease-species representations [59]. Pipelines with AI have been utilized to discover disulfiram, metformin, and thioridazine, among others, as new positions due to their repositioning [60].

4.9.3 Success stories and challenges

There are also clinical indications of repurposed agents, such as metformin (antidiabetic) for metabolite targeting, disulfiram (anti-alcoholism agent) for proteasome inhibition, and thioridazine (antipsychotic) for inducing apoptosis in leukemia [61].

Nonetheless, repurposing has issues such as dose optimization, off-target liabilities, and intellectual property limitations. Computational screening can help prevent these risks through predictive profiling, docking validation, and network-level profiling [62].

4.10 INTEGRATED WORKFLOW: A COHERENT COMPUTATIONAL PIPELINE

Integrated pipelines that consist of several layers of computations to guarantee reliability and translational potential are the best cancer drug discovery workflows [63].

Typical Workflow:

Target Identification: either through omics or through network analysis.
Structure modeling: experimental or predicted (e.g., AlphaFold).
Binding site analysis: pocket and hotspot mapping.

Virtual Screening: structure- and ligand-based docking.
Refinement and Rescoring: with the help of free energy and MD simulations.
ADMET, Toxicity Filtering: separate the wheat from the chaff.
Network Validation: bridging hits to disease-relevant networks.
Artificial Intelligence Optimization: perfect or create better analogs.
Drug Repurposing Overlapping: cross-reference current pharmacophores.
Experimental Feedback: Train models and establish top hits, and retrain models iteratively.

4.10.1 Case study: End-to-end computational pipeline in cancer drug discovery

A recent computational model combined HR-LC-MS profiling, molecular docking, MD simulations, ADMET screening, and network pharmacology to discover new phytochemicals against leukemia cell survival pathways [64].

Hybrid pipelines like these can reduce the attrition rate of such a process because they combine chemical diversity and computational rigor with biological relevance. They are an ideal example of how AI-assisted feedback-driven loops can simplify the process of cancer drug discovery from virtual hits to validated leads.

4.11 CHALLENGES, LIMITATIONS, AND FUTURE PERSPECTIVES

Computational oncology is confronted with acute problems even though rapid progress has been made.

4.11.1 Data quality, heterogeneity, and scarcity

The importance of AI models relies on biased datasets that are curated. The experimental protocols and missing metadata are varied, and there is a lack of class imbalance, which constrains reproducibility [65]. Standard data stores and benchmark programs (e.g., CACHE, Target2035) are used to deal with these problems.

4.11.2 Interpretability and explainability of a model

Although DL yields enhanced precision in predictions, it cannot provide a way of understanding its mechanism in a way that is readily regulated and accepted. New types of explainable AI (XAI) methods, including attention maps and feature attribution, make the decisions of models more transparent [66].

4.11.3 Applicability to clinical scenarios

The translation of in silico outcomes to clinical outcomes is still a challenge because of the tumor heterogeneity and complexity of the microenvironment. This gap in translation to patient-derived organoids and xenografts is filled with prospective validation [67].

4.11.4 Integration, interoperability, and reproducibility

To interface with a variety of tools and databases, standards and FAIR (Findable, Accessible, Interoperable, Reusable) data principles [68] need to be interoperable. The attempts at pipeline automation and cloud-based collaborative systems are enhancing reproducibility.

4.11.5 Emerging opportunities

Quantum computing, federated learning, and autonomous discovery systems using AI, robotics, and molecular simulation are the future of computational oncology [69]. These inventions are likely to bring forth unexplored chemical space and customized treatment plans.

The current developments in computational oncology and data-based biomedical studies will keep transforming contemporary cancer therapeutics. Massive genome scans have shown the occurrence and pathogenicity of specimens of germline variants in numerous cancers, giving a basis of knowledge on inherited cancer risk and shaping precision medicine strategies [70]. With the increasing volume and complexity of this type of multi-omics data, becoming a necessity to maintain reproducibility, transparency, and long-term usefulness of biomedical data resources, it is now necessary to comply with standard principles of data stewardship, including the FAIR (Findable, Accessible, Interoperable, Reusable) guidelines [71]. In addition to these breakthroughs, the development of quantum algorithms introduces transformative computing, allowing the simulation of molecular systems with high accuracy and introducing a new opportunity in terms of quantum-enabled drug discovery and modeling of chemical processes at new levels [72]. All these factors point to the possibility of a future where genome understanding, well-controlled data management, and quantum computing come to her convergence to enhance faster next-generation scientific research and treatment creation.

4.12 CONCLUSION

Computational strategies have revolutionized cancer drug discovery as a combination of structural biology, chemoinformatics, molecular simulation, and AI in a data-driven approach [73]. These computational fields give researchers the possibility to quickly search through an enormous amount of chemicals and find promising candidates much quicker than before. They can also be used to optimize lead compounds using repeated computational methods before there is any need for experimental validation [74]. Additionally, ML models can be used to predict ADMET risks and focus on druggable binding pockets as well as identify unexpected DTIs. This significantly saves a lot of money in the lab and in early phases of phase 1 discovery [75]. Together, these methods provide a solid foundation for rational drug design and systematic drug repurposing of already existing drugs.

The combination of multi-omics data (genomics, transcriptomics, proteomics, and metabolomics), network pharmacology, and natural product databases will stimulate the understanding of mechanisms and chemical diversity [76]. By identifying bioactive compounds mapped onto disease-related signaling networks, researchers can sort out the multi-target effects of both phytochemicals and synthetic molecules [77]. This way, they can develop polypharmacological agents that have an effect on complex cancer pathways. This integration also allows for an increase in knowledge about tumor evolution, tumor resistance mechanism, and tumor dependency in microenvironment [78]. Despite certain challenges, particularly in terms of data quality and model clarity and also in terms of data translation from these models to clinical results, the combination of AI, multi-omics analysis, and scalable high-performance computing is leading to a new era of Precision Therapies in oncology [79,80].

Looking into the future, the most important change is likely to come from adaptive pipelines, real-time, self-learning computational frameworks that continuously obtain better predictive accuracy by direct interaction with the experimental results. In this closed-loop system, the models of AI will improve structural predictions, propose optimized analogs, and adjust the pharmacological hypotheses based on new in vitro, in vivo, or clinical data. This dynamic relationship between computation and experimentation will not only be able to shorten the time needed for discoveries but also a personalized design of cancer drugs based on specific patient molecular profiles. Ultimately, these autonomous, evolving systems are the future of next-generation oncology drug discovery where prediction, validation and optimization of a process are operating in a seamless, intelligent cycle.

REFERENCES

1. Justus, C. R., Leffler, N., Ruiz-Echevarria, M., & Yang, L. V. (2021). In vitro cell migration and invasion assays. *Cancer Metastasis Reviews*, *40*(1), 1–27.
2. Singh, N., Kumar, S., & Verma, R. (2023). Computational insights in oncology drug development: Current status and future directions. *Computational Biology and Chemistry*, *104*, 107866.

3. Gangwal, R., & Lavecchia, A. (2025). Advances in cheminformatics-driven cancer drug discovery. *Drug Discovery Today, 30*(4), 102123.
4. Amorim, R., Dantas, R. A., & Pita, J. S. (2024). Computational strategies for predicting drug–target interactions in cancer therapy. *Briefings in Bioinformatics, 25*(2), bbae050.
5. Satheeskumar, L. (2025). Deep learning for precision oncology: Predictive and mechanistic modeling. *Frontiers in Artificial Intelligence, 8*, 1442109.
6. Mouvanal, S., & Basalingappa, K. M. (2024a). *Phytochemical interventions in acute myeloid leukemia: Mechanistic insights into apoptosis, oxidative stress modulation, and signaling pathway regulation.* Chinese Herbal Medicines.
7. Lavecchia, A. (2020). Machine-learning approaches in drug discovery: Methods and applications. *Drug Discovery Today, 25*(10), 1704–1712.
8. Koyuturk, M., & Wang, C. (2022). Precision oncology enabled by next-generation computational genomics. *Nature Reviews Clinical Oncology, 19*(9), 573–589.
9. Lill, M. A. (2020). Virtual screening in drug design. *Wiley Interdisciplinary Reviews: Computational Molecular Science, 10*(3), e1467.
10. Lawson, C. L., & Berman, H. (2021). Protein Data Bank: Current status and future plans. *Nucleic Acids Research, 49*(D1), D458–D465.
11. Baek, M., et al. (2021). Accurate prediction of protein structures and interactions using RoseTTAFold. *Science, 373*(6557), 871–876.
12. Trott, O., & Olson, A. J. (2020). AutoDock Vina 2.0 improvements. *Journal of Chemical Theory and Computation, 16*(1), 1–7.
13. Cuzzolin, A., et al. (2021). DockBench 2.0: Benchmarking of docking workflows. *Journal of Chemical Information and Modeling, 61*, 4081–4091.
14. Kozakov, D., et al. (2020). FTMap for binding hotspot identification. *Nature Protocols, 15*, 291–311.
15. Smith, J. A., & Johnson, T. (2022). Hotspot targeting of kinases and DNA-damage proteins in oncology. *Trends in Pharmacological Sciences, 43*(4), 289–304.
16. Gromiha, M. M., & Ou, Y. Y. (2021). Current developments in ligand-based drug design. *Current Opinion in Structural Biology, 71*, 153–160.
17. Wolber, G., & Langer, T. (2021). LigandScout 4.4: Recognition of pharmacophore features. *Journal of Chemical Information and Modeling, 61*(9), 4339–4350.
18. Kotev, M., et al. (2022). Pharmacophore-driven natural product screening. *Computational and Structural Biotechnology Journal, 20*, 1321–1331.
19. Cherkasov, A., et al. (2021). QSAR modeling for drug discovery. *Chemical Reviews, 121*(16), 10448–10525.
20. Baskin, I. I., & Varnek, A. (2020). Development of QSAR using ML methods. *Molecular Informatics, 39*(1), 1900090.
21. Shao, L., & Zheng, W. (2022). Natural product analog generation through QSAR-driven scaffold hopping. *Phytomedicine, 105*, 154378.
22. Hollingsworth, S. A., & Dror, R. O. (2020). Molecular dynamics in drug discovery. *Neuron, 107*(1), 77–91.
23. Lee, J., et al. (2020). CHARMM-GUI updates for MD simulations. *Journal of Chemical Physics, 152*(3), 034101.
24. Maiti, R., & Van Der Spoel, D. (2021). MD for cancer target refinement. *Journal of Chemical Information and Modeling, 61*, 4034–4047.

25. Laio, A., & Gervasio, F. L. (2022). Metadynamics in drug discovery. *Nature Reviews Chemistry, 6*, 559–575.
26. Killian, B. J., et al. (2021). MM/PBSA and FEP in ligand ranking. *Journal of Chemical Theory and Computation, 17*(4), 2353–2369.
27. Souza, P. C. T., et al. (2021). Martini 3 coarse-grained MD. *Nature Methods, 18*, 382–388.
28. Lavecchia, A. (2020). Virtual screening strategies in drug discovery. *Expert Opinion on Drug Discovery, 15*, 1–15.
29. Gentile, F., et al. (2020). HTVS with AI-based docking. *Journal of Chemical Information and Modeling, 60*(9), 4190–4200.
30. Mysinger, M. M., & Kuntz, I. D. (2021). Screening library design with ML rescoring. *Journal of Medicinal Chemistry, 64*, 180–195.
31. Sorokina, M., & Steinbeck, C. (2020). COCONUT natural product database. *Journal of Cheminformatics, 12*, 20.
32. Lv, H., et al. (2022). Network pharmacology for natural products. *Briefings in Bioinformatics, 23*(1), bbab540.
33. Chen, Z., et al. (2021). In silico–in vitro integrated screening. *Frontiers in Pharmacology, 12*, 671500.
34. Lo, Y., & Tsai, F. (2023). LC-MS/MS–docking hybrid approaches. *Analytica Chimica Acta, 1264*, 341239.
35. Hopkins, A. L. (2020). Network pharmacology. *Nature Chemical Biology, 16*, 556–565.
36. Szklarczyk, D., et al. (2021). STRING v11.5. *Nucleic Acids Research, 49*, D605–D612.
37. Zhou, Y., et al. (2023). Network-based target ranking in oncology. *Briefings in Bioinformatics, 24*(2), bbad056.
38. Anighoro, A., Bajorath, J., & Rastelli, G. (2020). Polypharmacology in drug design. *Drug Discovery Today, 25*, 15–24.
39. Gao, J., et al. (2022). AI-driven kinase inhibitor profiling. *Nature Machine Intelligence, 4*, 665–676.
40. Chuang, E., et al. (2021). Systems biology in cancer modeling. *Nature Reviews Cancer, 21*, 718–732.
41. Huang, C., et al. (2022). Systems-level evaluation of PI3K/MAPK/DDR networks. *Cell Systems, 13*(4), 272–287.
42. Vamathevan, J., et al. (2020). Applications of ML to drug discovery. *Nature Reviews Drug Discovery, 19*, 463–477.
43. Stokes, J. M., et al. (2020). GNNs for molecular prediction. *Cell, 180*(5), 1153–1169.
44. Wang, Z., et al. (2024). CandidateDrug4Cancer: Benchmark for DTI. *Nucleic Acids Research, 52*(D1), D392–D401.
45. He, T., et al. (2021). Hybrid embeddings for affinity prediction. *Bioinformatics, 37*, 440–448.
46. Blanchard, A. E., et al. (2021). Generative models for molecular design. *Nature Communications, 12*, 1403.
47. Jeon, Y., et al. (2022). RL-optimized de novo inhibitors. *Journal of Medicinal Chemistry, 65*, 9305–9321.
48. Kuenzi, B. M., et al. (2020). Predicting drug sensitivity with DL. *Nature Cancer, 1*, 820–832.

49. Li, Y., et al. (2023). Quantum neural networks in drug response prediction. *NPJ Digital Medicine, 6*, 55.
50. Sheller, M. J., et al. (2020). Federated learning in multicenter oncology. *Scientific Reports, 10*(1), 125.
51. Daina, A., & Zoete, V. (2021). SwissADME updates. *Nucleic Acids Research, 49*, W276–W279.
52. Pires, D. E., et al. (2021). pkCSM extensions. *Journal of Chemical Information and Modeling, 61*, 4869–4876.
53. Hsieh, J. H., et al. (2022). AI-driven toxicogenomics. *Toxicological Sciences, 186*(2), 209–221.
54. Grime, K., et al. (2023). PBPK–ML integrated toxicity prediction. *Drug Metabolism Reviews, 55*(1), 33–58.
55. Pushpakom, S., et al. (2020). Drug repurposing landscape. *Nature Reviews Drug Discovery, 19*, 41–58.
56. Subramanian, A., et al. (2020). Next-gen connectivity map. *Cell, 180*(2), 524–541.
57. Lee, J. S., et al. (2022). Network-integrated repurposing. *Nature Communications, 13*, 5618.
58. Jin, W., et al. (2021). AI-driven drug–disease link prediction. *Nature Communications, 12*, 5274.
59. Fang, X., et al. (2023). TxGNN for zero-shot repurposing. *Nature Machine Intelligence, 5*, 777–789.
60. Sonawane, A. R., et al. (2022). AI discovery of repurposable oncology agents. *NPJ Systems Biology and Applications, 8*, 38.
61. Lin, J., et al. (2021). Repurposed agents in leukemias. *Frontiers in Oncology, 11*, 640474.
62. Yang, Y., et al. (2023). Computational profiling to mitigate repurposing risks. *Drug Discovery Today, 28*(10), 103686.
63. Wang, H., et al. (2021). Integrated ML–MD–ADMET workflows. *Briefings in Bioinformatics, 22*(3), bbaa250.
64. Patel, J., et al. (2023). Hybrid pipeline for phytochemical-leukemia inhibitors. *Journal of Ethnopharmacology, 301*, 115829.
65. Fourches, D., et al. (2020). Data quality in ML models. *Journal of Chemical Information and Modeling, 60*, 4200–4210.
66. Jiménez-Luna, J., et al. (2021). XAI for molecular ML. *Nature Reviews Chemistry, 5*, 148–167.
67. Choi, J., et al. (2023). Translating computational hits via organoids. *Nature Biomedical Engineering, 7*, 376–389.
68. Wilkinson, M. D., et al. (2021). FAIR data guidelines 2021 update. *Scientific Data, 8*, 227.
69. Bian, Y., et al. (2022). AI-robotics autonomous discovery. *Nature Chemistry, 14*, 112–122.
70. Hu, C., et al. (2021). Germline variants in cancer genomics. *Nature Communi cations, 12*, 1438.
71. Sansone, S. A., et al. (2020). FAIR data stewardship. *Nature Biotechnology, 38*, 355–364.
72. Cao, Y., et al. (2022). Quantum computing for molecular simulation. *Chemical Reviews, 122*(5), 4497–4559.

73. Vamathevan, J., Clark, D., Czodrowski, P., Dunham, I., Ferran, E., Lee, G., Li, B., Madabhushi, A., Shah, P., Spitzer, M., & Watson, J. (2020). Applications of machine learning in drug discovery and development. *Nature Reviews Drug Discovery, 19*(7), 463–477.
74. Hollingsworth, S. A., & Dror, R. O. (2020). Molecular dynamics simulation for all. *Neuron, 107*(1), 77–91.
75. Daina, A., & Zoete, V. (2021). A BOILED-Egg to predict gastrointestinal absorption and brain penetration of small molecules. *Scientific Reports, 11*, 22543.
76. Barabási, A.-L., Menichetti, G., & Loscalzo, J. (2020). The unmapped chemical complexity of our diet. *Nature Food, 1*, 33–37.
77. Fourches, D., Muratov, E., & Tropsha, A. (2020). Trust but verify: On the importance of chemical structure curation in cheminformatics and QSAR modeling. *Journal of Chemical Information and Modeling, 60*(10), 4200–4210.
78. Sheller, M. J., Reina, G. A., Edwards, B., Martin, J., & Bakas, S. (2020). Multi-institutional deep learning modeling without sharing patient data: A federated learning approach. *Scientific Reports, 10*, 125.
79. Baek, M., et al. (2021). Accurate prediction of protein structures and interactions using RoseTTAFold. *Science, 373*(6557), 871–876.
80. Cao, Y., Romero, J., Olson, J. P., Degroote, M., Johnson, P. D., Kieferová, M., Kivlichan, I. D., Menke, T., Peropadre, B., Sawaya, N. P. D., & Aspuru-Guzik, A. (2022). Quantum chemistry in the age of quantum computing. *Chemical Reviews, 122*(5), 4497–4559.

Chapter 5

Structure-based approaches for infectious disease therapeutics

Kuralayanapalya Puttahonnappa Suresh[1], *Varsha Ramesh*[1], *and Nagaraju Chaithra*[2]

[1]Disease Informatics Lab, ICAR-National Institute of Veterinary Epidemiology and Disease Informatics (NIVEDI), Bengaluru, Karnataka, India

[2]Division of Medical Statistics, School of Life Sciences, JSS Academy of Higher Education & Research, Mysuru, Karnataka, India

5.1 INTRODUCTION

In the perpetual arms race between human and infectious pathogens, three-dimensional molecular insights into molecular architecture have emerged as one of the most powerful weapons in our therapeutic arsenal. Infectious diseases remain a global threat, driven by rapidly evolving pathogens and rising antimicrobial resistance. Traditional drug discovery, often reliant on trial and error, struggles to keep pace with these challenges. The field of structure-based drug discovery (SBDD) has fundamentally transformed how we approach infectious diseases, shifting from empirical observations to rational, molecular-level interventions [1]. The COVID-19 pandemic served as a stark reminder of our vulnerability to infectious threats while simultaneously showcasing the unprecedented power of modern structural biology. Structure-based approaches to infectious disease therapeutics operate on a deceptively simple principle: if we can visualize the molecular machinery that pathogens use to infect, replicate, and survive within their hosts, we can design precise interventions to disrupt these processes [2].

5.2 HISTORICAL OVERVIEW OF STRUCTURE-BASED APPROACHES

The evolution of SBDD has unfolded through a series of transformative milestones that have continually reshaped the field of therapeutics. The story begins in 1928, when Alexander Fleming's return from a brief vacation led to one of the most serendipitous findings in medical history. He noticed that a culture of *Staphylococcus aureus* had been lysed by a contaminating

DOI: 10.1201/9781003672760-5

mold, *Penicillium notatum*, which secreted a substance later identified as penicillin. Over a decade later, Howard Florey and Ernst Chain developed practical methods for its large-scale production, unveiling penicillin's therapeutic potential and forever changing the treatment of bacterial infections. Around the same time, the 1930s brought another leap forward with the discovery of sulfonamides. Originating from research on chemical dyes, German scientists, including Gerhard Domagk, observed that certain dye compounds selectively bound to bacterial cells. This insight paved the way for the creation of the first synthetic bacteriostatic drugs. These discoveries, while partially accidental, required analytical thinking and systematic experimentation to transform chance observations into life-saving therapies [3]. As structural biology advanced, the early dominance of X-ray crystallography offered the first atomic-level perspectives of pathogen proteins, laying the groundwork for rational drug design. The introduction of cryo-electron microscopy in the mid-20th century further expanded our understanding, revealing large, flexible, and membrane-bound molecular assemblies in unprecedented detail. By the 1980s and 1990s, structural analyses of bacterial β-lactamases and viral enzymes, such as HIV-1 protease, guided the rational development of inhibitors that fit their targets with exquisite precision [4]. Most recently, the integration of artificial intelligence (AI) and machine learning (ML) has begun to redefine not only how we determine molecular structures but also how we interpret their dynamics, predict their functional behavior, and design novel therapeutics. What began with accidental discoveries and painstaking crystallography has evolved into an era where data-driven intelligence accelerates our understanding of molecular systems and the design of life-saving interventions [5].

5.3 FOUNDATIONS OF STRUCTURE-BASED APPROACHES IN INFECTIOUS DISEASES

5.3.1 The structural basis of pathogen vulnerability

Infectious pathogens, whether viral, bacterial, fungal, or parasitic, must accomplish several fundamental tasks to successfully establish infection and propagate within their hosts. To ensure survival and transmission, these organisms must gain entry into host cells or tissues, replicate their genetic material, synthesize vital biomolecules, evade immune surveillance, and ultimately spread to new hosts [6]. Each of these processes depends on specific protein–protein interactions, enzyme-catalyzed reactions, and structural transformations that together form the molecular foundation of infection. These same mechanisms also provide potential points of vulnerability, precise targets for therapeutic intervention. Identifying druggable targets within pathogen proteomes requires a deep understanding of both the biological indispensability of specific proteins and their structural

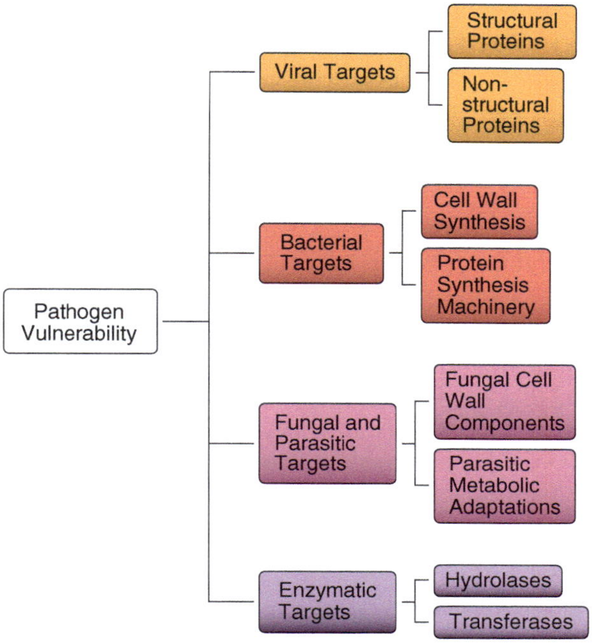

Figure 5.1 Classification of macromolecular targets contributing to pathogen vulnerability. Each category highlights critical molecular processes or proteins that can be exploited for therapeutic intervention.

accessibility to small-molecule inhibitors. The influential work of Hopkins and Groom introduced the concept of the "druggable genome," proposing that only about 10%–14% of the human proteome comprises proteins that can be effectively targeted by small molecules [7]. In the case of pathogenic organisms, this fraction is likely even smaller, reflecting their compact and highly optimized genomes. As a result, the process of target selection becomes exceptionally critical in pathogen-focused drug discovery. Broadly, the macromolecular targets in pathogens can be categorized according to their molecular function, the pathogen's taxonomic group, or the specific interactions they mediate at the host–pathogen interface (Figure 5.1).

5.3.1.1 *Viral targets*

Viral pathogens present a unique set of structural targets due to their obligate dependence on host cellular machinery for replication, while also maintaining essential virus-specific proteins required for infection and propagation. The limited coding capacity of viral genomes creates a concentrated set of potential targets, each often serving multiple essential functions.

Structural Proteins: These represent the most visible class of viral targets, including capsid proteins responsible for genome packaging and particle stability, envelope proteins that mediate cellular entry, and matrix proteins that coordinate viral assembly [8].

Non-Structural Proteins: They encompass the enzymatic machinery essential for viral replication, including RNA-dependent RNA polymerases, reverse transcriptase, and proteases.

5.3.1.2 Bacterial targets

Bacterial pathogens offer numerous opportunities for selective therapeutic intervention due to fundamental differences in cellular organization and metabolism compared to eukaryotic hosts. The prokaryotic nature of bacterial cells provides multiple pathogen-specific targets that have been successfully exploited in antibiotic development.

Cell Wall Synthesis: represents perhaps the most successful class of bacterial targets, with β-lactam antibiotics targeting peptidoglycan synthesis enzymes that have no counterparts in human cells [9]. Structural elucidation of penicillin-binding proteins has illuminated the molecular underpinnings of both β-lactam action and emerging resistance mechanisms, paving the way for the design of β-lactamase inhibitors and newer generations of β-lactam antibiotics with broadened activity spectra.

Protein Synthesis: represents a key target for antibacterial action, as the structural distinction between bacterial (70S) and eukaryotic (80S) ribosomes allows for selective inhibition. Recent cryo-EM studies have further clarified how antibiotics interact with bacterial ribosomes, offering valuable insights into their binding mechanisms and the emergence of resistance development [10].

5.3.1.3 Fungal and parasitic targets

Eukaryotic pathogens pose greater challenges for selective drug development because of their close evolutionary similarity to humans [11]. Nevertheless, structural biology has identified subtle yet exploitable differences that enable the creation of targeted antifungal and antiparasitic therapies.

Fungal Cell Wall Components: Fungal cell wall components such as β-1,3-glucan synthase are absent in mammalian cells, providing unique targets for drugs like the echinocandins. Likewise, differences in the ergosterol biosynthesis pathway, particularly in fungal cytochrome P450 enzymes, have been harnessed in the design of azole antifungals, which selectively disrupt fungal membrane synthesis [12].

Parasitic Metabolic Adaptations: often reflect differences in the ergosterol biosynthesis pathway. Particularly, fungal cytochrome P450 enzymes have been harnessed in the design of azole antifungals, which selectively disrupt fungal membrane synthesis [13].

5.3.1.4 Enzymatic targets

Enzymes remain among the most accessible and well-characterized drug targets due to their defined active sites and predictable catalytic mechanisms [14]. Grouping enzymes by their reaction type helps identify their druggability and potential for selective inhibition. Hydrolases, such as proteases, nucleases, and β-lactamases, are proven therapeutic targets because of their distinct and deep active sites. Transferases, including kinases and methyltransferases, also provide promising opportunities as their cofactor binding pockets can be strategically targeted to design effective inhibitors.

5.3.2 The integration of structural biology with medicinal chemistry

The translation of structural insights into effective therapeutics requires seamless integration of structural biology with medicinal chemistry. This integration has evolved from simple structure–activity relationships (SARs) to sophisticated computational approaches that can predict and optimize molecular properties [15]. Modern SBDD employs iterative cycles of structural determination, computational modeling, chemical synthesis, and biological evaluation. Each cycle refines our understanding of SARs and guides the optimization of lead compounds. High-resolution structural data enables medicinal chemists to make rational modifications to improve potency, selectivity, pharmacokinetics, and other drug-like properties.

The concept of "structure-enabled drug design" has emerged as a more comprehensive approach that recognizes the dynamic nature of both targets and ligands [16]. Rather than treating protein structures as static entities, this approach considers conformational flexibility, binding kinetics, and the thermodynamic basis of molecular recognition. These considerations have become increasingly important as our understanding of protein dynamics has grown and computational tools have evolved to model these complex phenomena.

5.4 TECHNOLOGICAL FOUNDATIONS: FROM X-RAYS TO AI

5.4.1 X-ray crystallography: The gold standard

X-ray crystallography continues to provide the majority of high-resolution structural information for infectious disease targets, contributing approximately 85% of all pathogen protein structures in the Protein Data Bank [17]. The technique's ability to provide atomic-level detail has made it indispensable for understanding binding sites, designing inhibitors, and optimizing drug–target interactions (Figure 5.2). The application of crystallography to

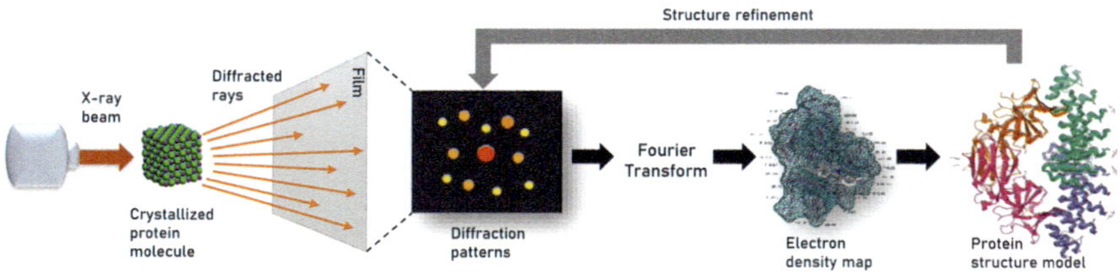

Figure 5.2 Workflow of X-ray crystallography, depicting the stepwise process from protein purification to structure prediction.

infectious disease research has required specialized approaches to handle the unique challenges presented by pathogen proteins. Many viral proteins are inherently unstable, existing as transient intermediates in complex replication cycles. Bacterial proteins may require specific cofactors or environmental conditions for proper folding. Parasitic enzymes often possess unique structural features that reflect their specialized metabolic requirements.

Modern crystallographic approaches to pathogen research employ high-throughput methods that can rapidly screen multiple protein constructs, crystallization conditions, and data collection strategies [18]. Fragment-based approaches have become particularly powerful when combined with crystallographic screening. Fragment libraries containing hundreds to thousands of small, drug-like molecules can be screened against pathogen targets using automated soaking and data collection procedures. The resulting crystal structures reveal binding modes and provide starting points for medicinal chemistry optimization, often leading to more efficient lead discovery than traditional high-throughput screening approaches [19].

5.4.2 Cryo-electron microscopy: revealing dynamic complexes

The "resolution revolution" in cryo-electron microscopy has transformed our ability to study large, dynamic protein complexes that are central to pathogen biology [20]. This technique has proven particularly valuable for characterizing viral particles, bacterial secretion systems, and host–pathogen interaction complexes that resist crystallographic analysis. Cryo-EM excels at visualizing macromolecular assemblies in states that closely approximate their native conformations. Viral capsids, envelope proteins, and replication complexes can be studied without the need for crystallization, revealing structural details that may be lost or distorted in crystal lattices. The technique's ability to resolve multiple conformational states from single datasets provides unprecedented insights into the dynamic processes that govern infection and replication.

The integration of cryo-EM with drug discovery efforts has led to new opportunities for structure-based design. Large viral complexes such as

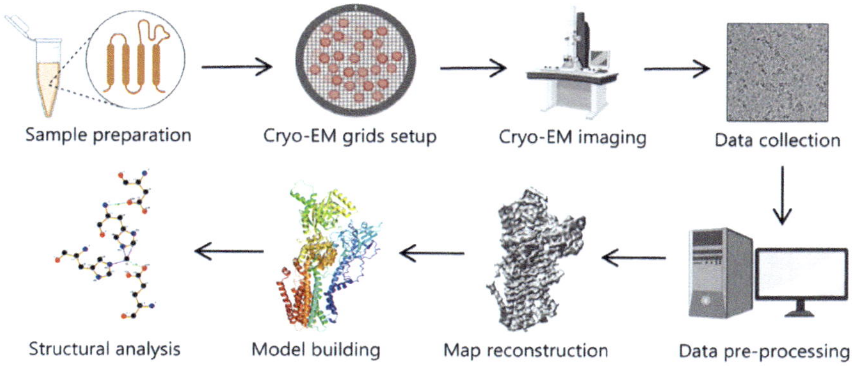

Figure 5.3 Schematic representation of cryo-electron microscopy (cryo-EM), showing the workflow from sample preparation to data processing.

RNA-dependent RNA polymerases can be visualized with bound inhibitors, revealing binding modes and allosteric effects that guide optimization efforts. Host–pathogen interaction complexes can be characterized to identify intervention points that disrupt essential processes without affecting normal host cell functions. Recent advances in cryo-EM methodology have expanded the technique's applicability to smaller proteins and complexes (Figure 5.3). Direct electron detectors, improved electron optics, and sophisticated image processing algorithms have pushed resolution limits below 2 Å for favorable samples [21]. These improvements have made cryo-EM competitive with crystallography for many applications while maintaining the advantages of studying proteins in more native-like environments.

5.4.3 Nuclear magnetic resonance spectroscopy

Nuclear magnetic resonance (NMR) spectroscopy provides unique insights into pathogen protein structure and dynamics in solution conditions that closely approximate physiological environments [22]. Unlike X-ray crystallography, which provides static snapshots, NMR can characterize protein flexibility, conformational exchange processes, and the dynamic states that are often critical for pathogen function. The technique has proven particularly valuable for studying intrinsically disordered viral proteins, which resist crystallization but play essential roles in viral replication and host cell manipulation. NMR-based fragment screening offers complementary advantages to crystallographic approaches, detecting weak binding interactions (millimolar affinities) and providing binding site information without requiring protein crystals. This capability is especially important for challenging infectious disease targets such as membrane proteins and conformationally dynamic enzymes. The technique's ability to monitor real-time

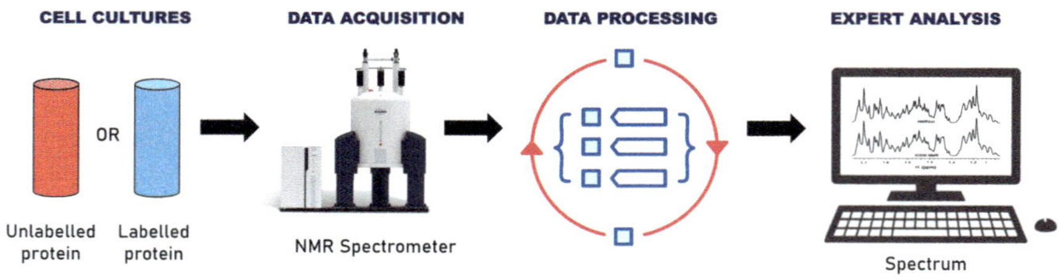

Figure 5.4 Workflow of nuclear magnetic resonance (NMR) spectroscopy, illustrating the stepwise process.

binding kinetics has revealed critical insights into drug mechanisms, allosteric regulation, and the molecular basis of antibiotic resistance in bacterial pathogens [23] (Figure 5.4).

5.4.4 Computational structural biology

The complexity of infectious disease mechanisms often requires integrative approaches that combine multiple structural techniques with biochemical, biophysical, and computational methods. The exponential growth in computational power and the development of sophisticated algorithms have made computationally intensive approaches routine components of drug discovery pipelines. Homology modeling remains the most widely used computational approach for generating structural models of pathogen proteins. Automated servers such as SWISS-MODEL, I-TASSER, and Phyre2 can generate structural models for entire pathogen proteomes, providing starting points for SBDD efforts. The quality of these models depends on the availability of suitable templates and the degree of sequence similarity between target and template proteins. Molecular dynamics simulations provide insights into protein flexibility, allosteric mechanisms, and the effects of mutations or chemical modifications [24]. These simulations can validate structural models, predict the effects of drug binding, and identify cryptic binding sites that are not apparent from static structures. Advanced simulation methods such as free energy perturbation calculations can predict the effects of chemical modifications on binding affinity, guiding lead optimization efforts [25]. Virtual screening approaches use computational models to search large chemical databases for compounds likely to bind to specific targets. These methods can rapidly evaluate millions of compounds, focusing experimental efforts on the most promising candidates.

5.5 THE AI REVOLUTION IN STRUCTURE-BASED DRUG DISCOVERY

5.5.1 Machine learning in protein structure prediction

The application of AI to protein structure prediction has achieved revolutionary breakthroughs that are transforming structure-based drug discovery (SBDD). DeepMind's AlphaFold system demonstrated that deep learning approaches could predict protein structures with near-experimental accuracy for many proteins [26]. This breakthrough is particularly significant for infectious disease research, as many pathogen proteins lack close homologs with known structures. AlphaFold's success stems from its ability to learn complex relationships between amino acid sequences and three-dimensional structures from the extensive database of experimentally determined structures. The system uses deep neural networks to predict inter-residue distances and angles, which are then used to construct three-dimensional models. The quality of these predictions often rivals experimental structures, providing immediate access to structural information that might take months or years to obtain experimentally.

The AlphaFold Protein Structure Database now contains structural predictions for over 200 million proteins, including complete proteomes for thousands of organisms [27]. This resource has particular value for infectious disease research, as it provides structural models for essential pathogen proteins that have resisted experimental characterization. These models can guide experimental structure determination efforts, identify potential binding sites, and enable virtual screening approaches. Beyond AlphaFold, numerous other AI-based structure prediction methods have emerged, each with specific strengths and applications. ChimeraX AlphaFold, ColabFold, and ESMFold offer different approaches to structure prediction, while specialized methods like AlphaFold2-multimer predict the structures of protein complexes [28]. The diversity of available methods allows researchers to choose approaches best suited to their specific applications and targets.

5.5.2 AI-driven drug design and optimization

ML approaches are revolutionizing how we design and optimize drug candidates, offering the potential to dramatically accelerate the drug discovery process while improving success rates [29]. These approaches can learn from vast databases of chemical and biological information to identify patterns and relationships that are not apparent to human researchers.

Generative models represent one of the most exciting applications of AI in drug design [30]. These models can generate novel chemical structures with desired properties, effectively exploring chemical space more efficiently than traditional approaches. Variational autoencoders, generative adversarial networks, and other deep learning architectures can learn the underlying patterns that define drug-like molecules and generate new candidates with optimized properties.

Reinforcement learning approaches can optimize molecular properties through iterative design cycles [31]. These methods treat drug design as a sequential decision-making problem, learning to make chemical modifications that improve desired properties while maintaining drug-like characteristics. The integration of structural information with reinforcement learning enables the design of molecules specifically optimized for particular binding sites.

Multi-objective optimization represents a critical challenge in drug design, as successful drugs must satisfy numerous criteria simultaneously: potency, selectivity, pharmacokinetics, safety, and manufacturability [32]. ML approaches can learn complex trade-offs between these properties and identify regions of chemical space that offer optimal combinations of desired characteristics.

5.5.2.1 The first complete AI-designed drug: rentosertib

The pharmaceutical industry achieved a historic milestone with *rentosertib* (formerly ISM001–055), the first drug whose target and design were discovered by modern generative AI. Developed by Insilico Medicine using their proprietary Pharma. AI platform, *rentosertib* represents the first fully generative AI drug to reach human clinical trials, specifically Phase II trials with patients. The discovery process began in 2020 and demonstrated the complete AI-driven drug development pipeline. *Rentosertib* is a first-in-class small molecule targeting TNIK (Traf2- and Nck-interacting kinase) and was designed using generative AI to treat idiopathic pulmonary fibrosis (IPF) [33] (Figure 5.5). *Rentosertib* is the first investigational drug in which both the biological target and the therapeutic compound were discovered using generative AI.

The clinical validation of this AI-driven approach came with promising results. On June 3, 2025, the industry's first proof-of-concept clinical validation of AI-driven drug discovery was published in Nature Medicine, with Insilico Medicine reporting promising safety and efficacy results from a Phase IIa trial of *rentosertib*. In silico's AI-designed *rentosertib* improved lung function by 98 mL in IPF patients, marking the first phase 2a success for a fully AI-generated drug [34].

This breakthrough demonstrates several key advantages of AI-driven drug discovery: significantly reduced timeline from target identification to clinical trials (accomplished in approximately 3 years versus the traditional 6–8 years), systematic exploration of novel target classes that might

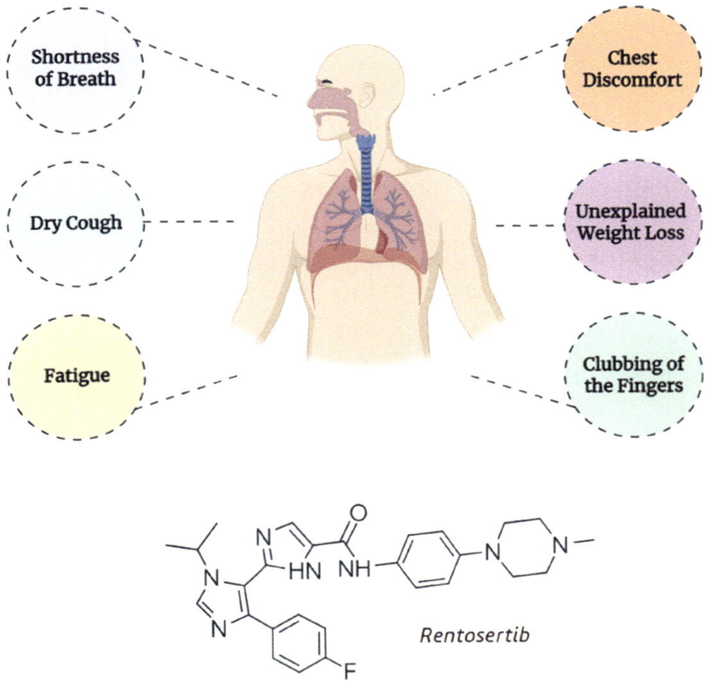

Figure 5.5 Clinical manifestations of idiopathic pulmonary fibrosis and molecular structure of rentosertib.

be overlooked by conventional approaches, and integrated optimization of multiple drug properties simultaneously. The success of *rentosertib* validates the potential of AI approaches to transform pharmaceutical research and provides a proof of concept for end-to-end AI-driven drug discovery pipelines.

5.5.3 AI in fragment-based drug discovery

Fragment-based drug discovery (FBDD) has become a powerful strategy for developing inhibitors against difficult therapeutic targets, and the integration of AI is advancing every stage of this process. ML algorithms enhance fragment library design, predict binding interactions, and support fragment-to-lead optimization. AI-driven library generation simultaneously considers chemical diversity, drug-likeness, synthetic feasibility, and predicted binding affinity, producing focused fragment sets that are more likely to yield high-quality hits for specific target classes [35]. The inclusion of structural information further enables the enrichment of libraries with fragments predisposed to bind certain structural motifs or active sites. Combining AI predictions with experimental fragment screening has resulted in hybrid workflows that leverage the strengths of both computational and empirical methods. ML models trained on experimental data can also refine the screening process by predicting likely outcomes, prioritizing fragments for testing, and identifying false positives or negatives, thereby improving the overall efficiency of FBDD pipelines.

5.5.4 Machine learning for resistance prediction and mitigation

The emergence of drug resistance remains a major obstacle in the development of effective anti-infective therapies. Pathogens evolve rapidly under selective pressure, generating resistant variants that diminish treatment efficacy. ML offers innovative approaches to predict and mitigate resistance mechanisms by analyzing the structural and functional consequences of genetic mutations. These predictive models assess parameters such as binding site conservation, protein stability, and functional constraints to determine which mutations can maintain protein function while reducing drug binding affinity [36].

AI-driven frameworks further support the design of next-generation therapeutics with improved resistance profiles [37]. By modeling the structural impacts of resistance-associated mutations, these methods can suggest chemical modifications that restore target binding or exploit alternative binding sites less susceptible to resistance. Additionally, AI facilitates combination therapy optimization by predicting synergistic drug pairs, dosing regimens, and treatment schedules that suppress the emergence of resistant strains while maintaining therapeutic potency.

5.6 QUANTUM COMPUTING AND DRUG DISCOVERY

Quantum computing is an emerging technology with the potential to transform aspects of SBDD. Because molecular interactions are governed by quantum mechanics, quantum computational approaches are particularly well-suited for modeling these systems and may outperform classical computers for certain calculations. Quantum chemistry simulations could improve predictions of binding affinity, selectivity, and chemical reactivity, thereby enhancing virtual screening, guiding lead optimization, and enabling more accurate forecasts of drug metabolism and toxicity [38]. Integrating quantum computing with AI may yield hybrid platforms that leverage the strengths of both technologies. Additionally, quantum approaches could be applied to drug–target interaction networks, whose multi-dimensional complexity may benefit from quantum optimization algorithms. These methods could identify optimal drug combinations or predict potential resistance pathways more efficiently than conventional computational strategies, opening new avenues for rational therapeutic design [39].

5.7 COMPREHENSIVE IN SILICO SBDD WORKFLOW

5.7.1 Workflow overview

5.7.1.1 Stage 1: Computational target identification and validation

The workflow begins with mining pathogen genomes retrieved from databases such as NCBI GenBank or Ensembl Bacteria to identify protein-coding sequences (Figure 5.6). ML classifiers trained on experimental essentiality data predict genes critical for pathogen survival, achieving 75%–85% accuracy by analyzing features including gene conservation, metabolic network position, and expression patterns during infection. Computational druggability assessment evaluates whether proteins possess binding pockets amenable to small-molecule intervention using sequence-based features (disorder content, similarity to known drug targets) and structure-based descriptors (pocket volume, depth, hydrophobicity) [40]. BLAST searches against human proteome databases identify potential off-target interactions, with proteins sharing >40% sequence identity requiring careful selectivity evaluation.

5.7.1.2 Stage 2: Structure acquisition and preparation

Target protein structures are obtained through PDB database searches for experimental structures (preferred resolution <2.5 Å) or generated computationally when experimental data are unavailable. Homology modeling creates structural models based on templates sharing >30% sequence identity using tools like MODELLER or SWISS-MODEL [41]. For targets lacking suitable templates, AlphaFold2 provides predicted structures often approaching experimental accuracy, with confidence metrics (pLDDT scores >70) indicating regions suitable for drug design [26]. Retrieved or predicted structures undergo preparation, including addition of missing atoms; protonation state assignment at physiological pH using tools like PROPKA; removal of non-essential crystal waters; and energy minimization to relieve steric clashes.

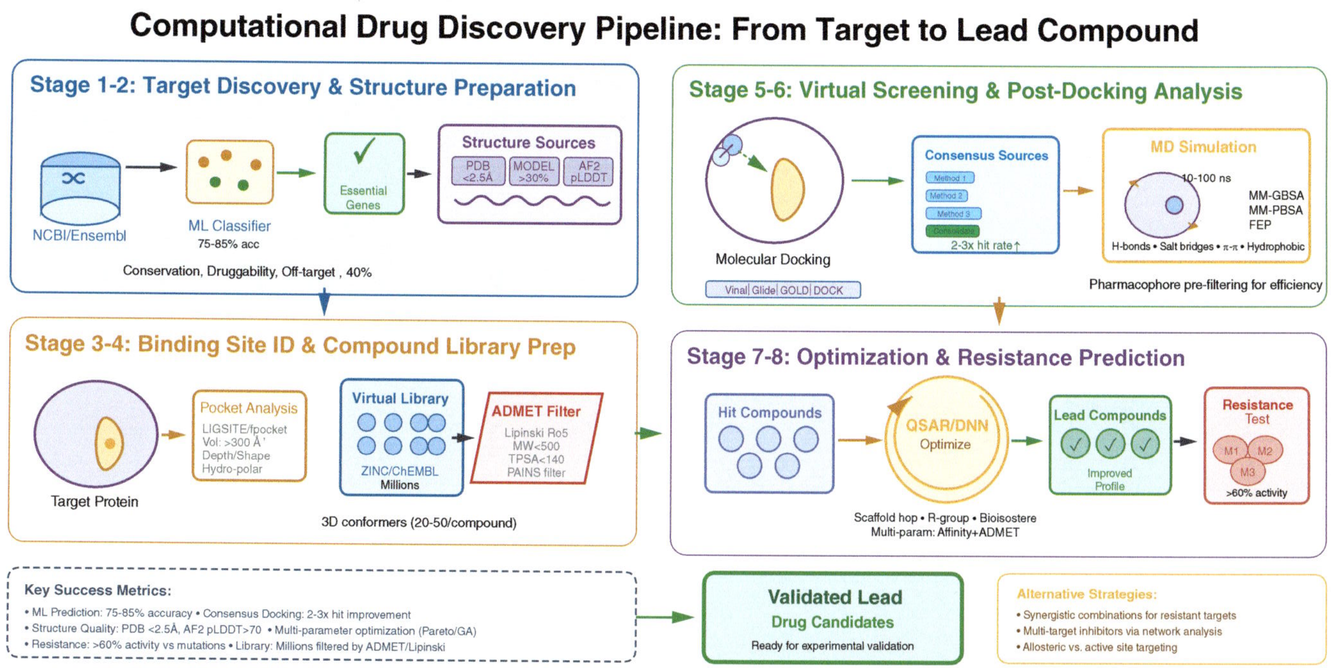

Figure 5.6 *In silico* structure-based drug design (SBDD) workflow.

5.7.1.3 Stage 3: Binding site identification and characterization

Geometric pocket detection algorithms analyze protein surface topology to identify potential binding sites using sphere-based approaches (LIGSITE), grid-based methods (fpocket), or alpha-shape techniques [42]. Computational probes representing different chemical functionalities are systematically placed throughout identified pockets to create interaction energy maps, revealing favorable binding regions. Identified sites are scored for druggability based on volume (>300 $Å^3$ preferred), depth, shape complementarity, hydrophobic-polar balance, and presence of specific interaction points. When multiple structures are available, comparative analysis distinguishes conserved obligatory binding sites from flexible allosteric regulatory sites.

5.7.1.4 Stage 4: Compound library preparation

Virtual compound libraries are assembled from commercial sources (ZINC, ChEMBL, PubChem), providing millions of purchasable compounds, specialized target-class libraries, or natural product databases. Computational ADMET filters remove compounds violating Lipinski's Rule of Five (MW<500, LogP<5, H-bond donors<5, acceptors<10) and additional criteria including TPSA<140 $Å^2$, rotatable bonds<10, and absence of pan-assay interference compound (PAINS) substructures [43]. Two-dimensional chemical structures are converted to three-dimensional coordinates with appropriate stereochemistry, generating multiple conformers (typically 20–50 per molecule) to sample accessible geometries using tools like OMEGA or RDKit. Compounds are protonated at physiological pH, and multiple tautomeric forms are generated for molecules with mobile protons [44].

5.7.1.5 Stage 5: Virtual screening by molecular docking

Molecular docking algorithms predict binding poses by systematically sampling ligand orientations and conformations within the defined binding site, scoring each pose using energy functions that estimate binding affinity [45]. Popular docking programs include AutoDock Vina (fast, empirical scoring), Glide (high accuracy, physics-based), GOLD (genetic algorithm search), and DOCK (grid-based scoring). Standard protocols involve defining the binding site (typically 10–15 Å from the pocket center), generating diverse ligand conformations, evaluating thousands to millions of poses, and clustering similar solutions [23]. Consensus scoring applies multiple independent scoring functions to evaluate docked poses, with compounds scoring well across multiple methods showing significantly improved hit rates (2–3× enhancement) compared to single scoring approaches [46].

Pharmacophore screening can pre-filter libraries based on essential molecular features (hydrogen bond donors/acceptors, hydrophobic centers, aromatic rings) before computationally expensive docking.

5.7.1.6 Stage 6: Post-docking analysis and refinement

Stage 6: Post-Docking Analysis and Refinement (2–4 weeks) Top-ranked compounds from virtual screening undergo molecular dynamics simulations (10–100 ns) in explicit solvent to assess binding pose stability, with unstable poses that rapidly dissociate being deprioritized [26]. More accurate binding affinity predictions employ methods including MM-GBSA/MM-PBSA (molecular mechanics with implicit solvent), free energy perturbation (FEP), or thermodynamic integration, which provide quantitative estimates correlating better with experimental data than docking scores alone [47]. Detailed interaction analysis identifies specific protein–ligand contacts, including hydrogen bonds (distance $<3.5\,\text{Å}$), salt bridges, π–π stacking, hydrophobic interactions, and water-mediated bridges. Selectivity predictions involve docking to off-target proteins (human homologs, related pathogen proteins) to identify compounds showing selective binding to the intended target.

5.7.1.7 Stage 7: Hit-to-lead optimization

Initial hit compounds undergo computational optimization to improve potency, selectivity, and drug-like properties. ML models trained on experimental SAR data predict how chemical modifications affect biological activity using QSAR models or deep neural networks [48]. Systematic in silico modifications include R-group enumeration (varying substituents), scaffold hopping (replacing core structures while maintaining key interactions), bioisosteric replacement (swapping functional groups), fragment growing (adding chemical groups), and fragment linking (connecting molecules binding adjacent sites). Multi-parameter optimization employs computational models to simultaneously optimize binding affinity, selectivity, solubility, permeability, metabolic stability, and toxicity risk using approaches like Pareto optimization or multi-objective genetic algorithms [49]. ADMET prediction models estimate pharmacokinetic properties (absorption, distribution, metabolism, excretion) and toxicity liabilities (hERG inhibition, hepatotoxicity, and mutagenicity) to guide selection of developable candidates.

5.7.1.8 Stage 8: Lead optimization and resistance prediction

Lead compounds undergo final refinement, including the introduction of known resistance mutations into structural models to predict compound activity against resistant variants. Docking and binding energy calculations identify compounds maintaining activity (>60% retention) against multiple resistance mutations, indicating robust resistance barriers [29]. ML models trained on clinical resistance data predict mutations likely to emerge under drug pressure, guiding the design of resistance-proof inhibitors that target highly conserved residues or make critical main-chain interactions less susceptible to mutation [41]. For targets prone to resistance development, computational methods can identify synergistic compound combinations or optimal multi-target strategies using network analysis and systems biology approaches [30].

5.8 CASE STUDIES: SUCCESS STORIES IN STRUCTURE-BASED ANTI-INFECTIVE DEVELOPMENT

5.8.1 HIV: The paradigm for structure-based antiviral development

The development of HIV therapeutics represents perhaps the most successful application of SBDD in the context of infectious diseases [50]. Detailed structural studies of key HIV enzymes, reverse transcriptase, protease, and integrase, have facilitated the creation of multiple classes of antiretroviral drugs, transforming HIV infection from a uniformly fatal disease into a manageable chronic condition.

Among these, HIV protease was the first viral enzyme targeted through systematic structure-based strategies. The enzyme's essential role in viral maturation and its unique structural features made it an attractive target for selective inhibition. Early structural studies revealed a homodimeric enzyme with a large, flexible active site capable of accommodating diverse peptide substrates [31].

The development of HIV protease inhibitors exemplified the power of iterative structure-based design. Initial peptidomimetic inhibitors were designed based on the structures of natural substrates but suffered from poor pharmacological properties. Subsequent generations incorporated non-peptidic scaffolds, improved selectivity, and enhanced resistance profiles, guided by detailed structural analyses of enzyme-inhibitor complexes. The emergence of drug resistance has posed significant challenges but also created opportunities for structure-based optimization of therapeutics. Structural analyses of resistance-associated mutations have elucidated the molecular mechanisms that reduce inhibitor binding and have informed the design of second- and third-generation inhibitors with enhanced resistance profiles. For example, *Darunavir*, a third-generation HIV protease inhibitor, was specifically engineered to retain efficacy against resistant variants by establishing extensive hydrogen-bonding networks within the active site. These interactions are difficult to disrupt through single amino acid substitutions.

5.8.2 SARS-CoV-2: Rapid response through structural biology

The COVID-19 pandemic highlighted the remarkable speed achievable with modern SBDD. Within weeks of the SARS-CoV-2 genome being sequenced, structural biology groups around the world had begun systematic characterization of viral proteins, accelerating the development of effective therapeutics.

The SARS-CoV-2 main protease (Mpro) quickly emerged as a key antiviral target due to its essential role in viral replication and the absence of closely related human homologs. Structural studies revealed a cysteine protease with distinctive substrate specificity, enabling selective inhibitor design. Its similarity to the SARS-CoV Mpro further facilitated rapid structure determination and drug development efforts.

Fragment-based approaches have played a pivotal role in identifying initial chemical starting points for Mpro inhibitors. At the Diamond Light Source XChem facility, over 1,000 fragments were screened against Mpro using automated crystallography, revealing multiple binding sites and diverse chemical scaffolds for optimization. These fragment hits provided valuable templates for medicinal chemistry programs across several pharmaceutical companies. The development of **Paxlovid (nirmatrelvir/ritonavir)** exemplifies the power of integrating structural biology with computational modeling and medicinal chemistry [51]. Pfizer's team leveraged structural insights from a compound initially developed for SARS-CoV, making targeted modifications to enhance potency, selectivity, and pharmacological properties. Continuous access to high-resolution crystal structures throughout the optimization process enabled rapid SAR analysis, facilitating swift and efficient lead optimization.

5.8.3 Tuberculosis: Addressing a global health challenge

Mycobacterium tuberculosis represents one of the most challenging infectious disease targets due to the bacterium's complex cell wall structure, slow growth rate, and ability to persist in latent forms within host cells. Structure-based approaches have been essential for understanding TB pathogenesis and identifying new therapeutic targets. The structural characterization of InhA, the target of the first-line anti-TB drug isoniazid, has revealed the molecular basis for the drug's mechanism of action and the structural consequences of resistance mutations [52]. This understanding has guided the development of next-generation InhA inhibitors with improved potency and resistance profiles, including compounds that maintain activity against isoniazid-resistant strains. Mycolic acid synthesis represents a unique feature of mycobacterial cell walls and has been extensively studied using structural approaches. The characterization of fatty acid synthase components, polyketide synthases, and modifying enzymes has revealed multiple potential targets for TB therapeutics. The structural complexity of these multi-domain enzymes has required integrative approaches combining X-ray crystallography, cryo-EM, and computational modeling. The development of bedaquiline, the first new anti-TB drug approved in over 40 years, was guided by structural studies of ATP synthase [53]. Structural characterization of the drug–target complex revealed the molecular basis for the compound's unique mechanism of action and provided insights into resistance mechanisms that have informed clinical development strategies.

5.9 REGULATORY CONSIDERATIONS AND QUALITY ASSURANCE

The integration of AI into SBDD raises new questions about validation, quality assurance, and regulatory acceptance [54]. Traditional regulatory frameworks were developed for experimentally determined structures and empirically optimized compounds, creating uncertainty about how AI-generated data should be evaluated and utilized. Validation standards for AI-predicted structures must consider both technical accuracy and biological relevance. While AlphaFold predictions may achieve high structural accuracy, their utility for drug discovery depends on the accuracy of binding site geometries, conformational flexibility, and allosteric networks that may not be fully captured in static predictions. Regulatory agencies are beginning to develop guidance documents for AI applications in drug discovery, but significant uncertainty remains about requirements for AI-designed therapeutics [55]. Early engagement with regulatory agencies and the development of standardized validation protocols will be essential for realizing the full potential of AI in infectious disease therapeutics.

High-throughput structural biology relies on robust quality control to ensure that results are reliable and reproducible. While automated data collection and analysis pipelines can handle vast amounts of information, they may overlook subtle errors or artifacts that expert human evaluation would catch. To address this, standardized validation protocols, automated error-detection tools, and regular benchmarking exercises are essential for maintaining data integrity [56]. Achieving a balance between speed and automation on one hand, and accuracy and reliability, on the other, is critical. Community standards and data-sharing practices further enhance reproducibility and trust in structural data. Initiatives such as the Protein Data Bank validation and remediation efforts, standardized deposition requirements, and open peer-review processes play a vital role in upholding high standards across the field of structural biology.

5.10 CONCLUSION

The convergence of structural biology, computational methods, and AI has transformed anti-infective drug discovery from a largely serendipitous endeavor into a rapid, rational, and knowledge-driven process. Achievements such as the swift development of SARS-CoV-2 therapeutics highlight the potential of modern structure-based approaches to respond to emerging threats. Looking ahead, the continued integration of AI, high-resolution structural data, and systems-level understanding promises to accelerate discovery, improve success rates, and uncover novel therapeutic strategies. Challenges remain, including drug resistance, difficult-to-study targets, and global health inequities, but interdisciplinary collaboration and technological innovation provide a robust framework for overcoming them. By combining molecular insight, computational power, and international cooperation, structure-based approaches position us to confront current and future infectious disease threats with unprecedented precision and speed.

REFERENCES

1. Blundell TL, Jhoti H, Abell C. High-throughput crystallography for lead discovery in drug design. *Nat Rev Drug Discov.* 2002;1(1):45–54. doi:10.1038/nrd706
2. Anderson AC. The process of structure-based drug design. *Chem Biol.* 2003;10(9):787–797. doi:10.1016/j.chembiol.2003.09.002
3. Ghosh AK, Gemma S. *Structure-Based Design of Drugs and Other Bioactive Molecules: Tools and Strategies.* 1st ed. Wiley; 2014. doi:10.1002/9783527665211
4. Renaud J, ed. *Structural Biology in Drug Discovery: Methods, Techniques, and Practices.* 1st ed. Wiley; 2020. doi:10.1002/9781118681121
5. Senior AW, Evans R, Jumper J, et al. Improved protein structure prediction using potentials from deep learning. *Nature.* 2020;577(7792):706–710. doi:10.1038/s41586-019-1923-7
6. Casadevall A, Pirofski L-A. Host-pathogen interactions: redefining the basic concepts of virulence and pathogenicity. *Infect Immun.* 1999;67(8):3703–3713. doi:10.1128/IAI.67.8.3703-3713.1999
7. Hopkins AL, Groom CR. The druggable genome. *Nat Rev Drug Discov.* 2002;1(9):727–730. doi:10.1038/nrd892
8. Harrison SC. Viral membrane fusion. *Nat Struct Mol Biol.* 2008;15(7):690–698. doi:10.1038/nsmb.1456
9. Bush K, Bradford PA. β-Lactams and β-lactamase inhibitors: an overview. *Cold Spring Harb Perspect Med.* 2016;6(8):a025247. doi:10.1101/cshperspect.a025247

10. Yusupov MM, Yusupova GZh, Baucom A, et al. Crystal structure of the ribosome at 5.5Å resolution. *Science.* 2001;292(5518):883–896. doi:10.1126/science.1060089
11. Fairlamb AH, Gow NAR, Matthews KR, Waters AP. Drug resistance in eukaryotic microorganisms. *Nat Microbiol.* 2016;1(7):16092. doi:10.1038/nmicrobiol.2016.92
12. Lepesheva GI, Waterman MR. Sterol 14α-demethylase cytochrome P450 (CYP51), a P450 in all biological kingdoms. *Biochim Biophys Acta BBA Gen Subj.* 2007;1770(3):467–477. doi:10.1016/j.bbagen.2006.07.018
13. Hyde JE. Exploring the folate pathway in Plasmodium falciparum. *Acta Trop.* 2005;94(3):191–206. doi:10.1016/j.actatropica.2005.04.002
14. Copeland RA. *Evaluation of Enzyme Inhibitors in Drug Discovery: A Guide for Medicinal Chemists and Pharmacologists.* 1st ed. Wiley; 2013. doi:10.1002/9781118540398
15. Hughes J, Rees S, Kalindjian S, Philpott K. Principles of early drug discovery. *Br J Pharmacol.* 2011;162(6):1239–1249. doi:10.1111/j.1476-5381.2010.01127.x
16. Sliwoski G, Kothiwale S, Meiler J, Lowe EW. Computational methods in drug discovery. *Pharmacol Rev.* 2014;66(1):334–395. doi:10.1124/pr.112.007336
17. Berman HM. The protein data bank. *Nucleic Acids Res.* 2000;28(1):235–242. doi:10.1093/nar/28.1.235
18. Stevens RC, Cherezov V, Katritch V, et al. The GPCR network: a large-scale collaboration to determine human GPCR structure and function. *Nat Rev Drug Discov.* 2013;12(1):25–34. doi:10.1038/nrd3859
19. Erlanson DA, Fesik SW, Hubbard RE, Jahnke W, Jhoti H. Twenty years on: the impact of fragments on drug discovery. *Nat Rev Drug Discov.* 2016;15(9):605–619. doi:10.1038/nrd.2016.109
20. Cheng Y. Single-particle cryo-EM at crystallographic resolution. *Cell.* 2015;161(3):450–457. doi:10.1016/j.cell.2015.03.049
21. Nakane T, Kotecha A, Sente A, et al. Single-particle cryo-EM at atomic resolution. *Nature.* 2020;587(7832):152–156. doi:10.1038/s41586-020-2829-0
22. Kleckner IR, Foster MP. An introduction to NMR-based approaches for measuring protein dynamics. *Biochim Biophys Acta BBA Proteins Proteomics.* 2011;1814(8):942–968. doi:10.1016/j.bbapap.2010.10.012
23. Palmer AG. NMR characterization of the dynamics of biomacromolecules. *Chem Rev.* 2004;104(8):3623–3640. doi:10.1021/cr030413t
24. Karplus M, McCammon JA. Molecular dynamics simulations of biomolecules. *Nat Struct Biol.* 2002;9(9):646–652. doi:10.1038/nsb0902-646
25. Cournia Z, Allen B, Sherman W. Relative binding free energy calculations in drug discovery: recent advances and practical considerations. *J Chem Inf Model.* 2017;57(12):2911–2937. doi:10.1021/acs.jcim.7b00564
26. Jumper J, Evans R, Pritzel A, et al. Highly accurate protein structure prediction with AlphaFold. *Nature.* 2021;596(7873):583–589. doi:10.1038/s41586-021-03819-2
27. Varadi M, Anyango S, Deshpande M, et al. AlphaFold protein structure database: massively expanding the structural coverage of protein-sequence space with high-accuracy models. *Nucleic Acids Res.* 2022;50(D1):D439–D444. doi:10.1093/nar/gkab1061

28. Evans R, O'Neill M, Pritzel A, et al. Protein complex prediction with AlphaFold-Multimer. Preprint posted online October 4, 2021. doi:10.1101/2021.10.04.463034
29. Iketani S, Forouhar F, Liu H, et al. Lead compounds for the development of SARS-CoV-2 3CL protease inhibitors. *Nat Commun.* 2021;12(1):2016. doi:10.1038/s41467-021-22362-2
30. Fromer M, Shifman JM. Tradeoff between stability and multispecificity in the design of promiscuous proteins. *PLoS Comput Biol.* 2009;5(12):e1000627. doi:10.1371/journal.pcbi.1000627
31. Wlodawer A, Miller M, Jaskólski M, et al. Conserved folding in retroviral proteases: crystal structure of synthetic HIV-1 protease. *Science.* 1989;245(4918):616–621. doi:10.1126/science.2548279
32. Jin Z, Du X, Xu Y, et al. Structure of Mpro from SARS-CoV-2 and discovery of its inhibitors. *Nature.* 2020;582(7811):289–293. doi:10.1038/s41586-020-2223-y
33. Xu Z, Ren F, Wang P, et al. A generative AI-discovered TNIK inhibitor for idiopathic pulmonary fibrosis: a randomized phase 2a trial. *Nat Med.* 2025;31(8):2602–2610. doi:10.1038/s41591-025-03743-2
34. Drug Target Review. First AI-designed drug, Rentosertib, officially named by USAN. March 14, 2025.
35. Yoo J, Jang W, Shin WH. From part to whole: AI-driven progress in fragment-based drug discovery. *Curr Opin Struct Biol.* 2025;91:102995. doi:10.1016/j.sbi.2025.102995
36. Palmer AC, Kishony R. Understanding, predicting and manipulating the genotypic evolution of antibiotic resistance. *Nat Rev Genet.* 2013;14(4):243–248. doi:10.1038/nrg3351
37. Stokes JM, Yang K, Swanson K, et al. A deep learning approach to antibiotic discovery. *Cell.* 2020;180(4):688–702.e13. doi:10.1016/j.cell.2020.01.021
38. Reiher M, Wiebe N, Svore KM, Wecker D, Troyer M. Elucidating reaction mechanisms on quantum computers. *Proc Natl Acad Sci.* 2017;114(29):7555–7560. doi:10.1073/pnas.1619152114
39. Cao Y, Romero J, Olson JP, et al. Quantum chemistry in the age of quantum computing. *Chem Rev.* 2019;119(19):10856–10915. doi:10.1021/acs.chemrev.8b00803
40. Cheng AC, Coleman RG, Smyth KT, et al. Structure-based maximal affinity model predicts small-molecule druggability. *Nat Biotechnol.* 2007;25(1):71–75. doi:10.1038/nbt1273
41. Waterhouse A, Bertoni M, Bienert S, et al. SWISS-MODEL: homology modelling of protein structures and complexes. *Nucleic Acids Res.* 2018;46(W1):W296–W303. doi:10.1093/nar/gky427
42. Le Guilloux V, Schmidtke P, Tuffery P. Fpocket: an open source platform for ligand pocket detection. *BMC Bioinformatics.* 2009;10(1):168. doi:10.1186/1471-2105-10-168
43. Lipinski CA, Lombardo F, Dominy BW, Feeney PJ. Experimental and computational approaches to estimate solubility and permeability in drug discovery and development settings. *Adv Drug Deliv Rev.* 1997;23(1–3):3–25. doi:10.1016/S0169-409X(96)00423-1

44. Martin YC. Let's not forget tautomers. *J Comput Aided Mol Des.* 2009;23(10):693. doi:10.1007/s10822-009-9303-2
45. Kitchen DB, Decornez H, Furr JR, Bajorath J. Docking and scoring in virtual screening for drug discovery: methods and applications. *Nat Rev Drug Discov.* 2004;3(11):935–949. doi:10.1038/nrd1549
46. Charifson PS, Corkery JJ, Murcko MA, Walters WP. Consensus scoring: a method for obtaining improved hit rates from docking databases of three-dimensional structures into proteins. *J Med Chem.* 1999;42(25):5100–5109. doi:10.1021/jm990352k
47. Genheden S, Ryde U. The MM/PBSA and MM/GBSA methods to estimate ligand-binding affinities. *Expert Opin Drug Discov.* 2015;10(5):449–461. doi:10.1517/17460441.2015.1032936
48. Cherkasov A, Muratov EN, Fourches D, et al. QSAR modeling: where have you been? Where are you going to? *J Med Chem.* 2014;57(12):4977–5010. doi:10.1021/jm4004285
49. Deb K, Pratap A, Agarwal S, Meyarivan T. A fast and elitist multiobjective genetic algorithm: NSGA-II. *IEEE Trans Evol Comput.* 2002;6(2):182–197. doi:10.1109/4235.996017
50. Flexner C. HIV-protease inhibitors. *N Engl J Med.* 1998;338(18):1281–1293. doi:10.1056/NEJM199804303381808
51. Lamb YN. Nirmatrelvir plus ritonavir: first approval. *Drugs.* 2022;82(5):585–591. doi:10.1007/s40265-022-01692-5
52. Sacchettini JC, Rubin EJ, Freundlich JS. Drugs versus bugs: in pursuit of the persistent predator Mycobacterium tuberculosis. *Nat Rev Microbiol.* 2008;6(1):41–52. doi:10.1038/nrmicro1816
53. Andries K, Verhasselt P, Guillemont J, et al. A diarylquinoline drug active on the ATP synthase of *Mycobacterium tuberculosis*. *Science.* 2005;307(5707):223–227. doi:10.1126/science.1106753
54. Topol EJ. High-performance medicine: the convergence of human and artificial intelligence. *Nat Med.* 2019;25(1):44–56. doi:10.1038/s41591-018-0300-7
55. Benjamens S, Dhunnoo P, Meskó B. The state of artificial intelligence-based FDA-approved medical devices and algorithms: an online database. *NPJ Digit Med.* 2020;3(1):118. doi:10.1038/s41746-020-00324-0
56. Gore S, Sanz García E, Hendrickx PMS, et al. Validation of structures in the protein data bank. *Structure.* 2017;25(12):1916–1927. doi:10.1016/j.str.2017.10.009

Chapter 6

In silico drug design for metabolic and cardiovascular diseases

Bhanumati Sarkar[1], *Sangram Polley*[2], *Sarthak Ranjan Sarkar*[3], *Shubhadeep Roychoudhury*[4], *Chandan Acharya*[5], *and Nithar Ranjan Madhu*[6*]

[1]Department of Botany, Acharya Prafulla Chandra College, North 24 Parganas, West Bengal, India

[2]Department of Life Sciences, Presidency University, Kolkata, West Bengal, India

[3]Department of Zoology, Bidhannagar College, Salt Lake, West Bengal, India

[4]Department of Life Science and Bioinformatics, Assam University, Silchar, Assam, India

[5]Department of Botany, Bajkul Milani Mahavidyalaya, Kismat Bajkul, West Bengal, India

[6]Department of Zoology, Acharya Prafulla Chandra College, North 24 Parganas, West Bengal, India

*Corresponding Author: nitharranjan.madhu@gmail.com

6.1 INTRODUCTION

Metabolic and cardiovascular diseases (CVDs) are one of the most significant public health challenges of the 21st century, accounting for a substantial and growing global disease burden. CVDs remain the leading cause of death worldwide. Approximately 31% of all deaths in India are caused by CVDs (Dorairaj et al., 2016). Metabolic disorders such as diabetes (Type 2), obesity, and hypercholesterolemia make the problem more complex. CVDs, like heart attacks, are a secondary harmful effect that arises as a result of long-term uncontrolled metabolic diseases, such as type 2 diabetes) (Acharya et al., 2023; Sarkar et al., 2023; Biswas et al., 2023; Roy et al., 2023; Sur et al., 2023; Sadhu et al., 2025). The multi-target drugs have drawn significant attention in recent years as a potential therapeutic solution to complex health conditions (Biswas et al., 2024; Das et al., 2024; Madhu et al., 2024; Roy et al., 2024). The classical "One Drug, One Target" approach struggles to address the complexity of metabolic and CVDs (Zheng et al., 2014). Therefore, the fundamental model shifts from a single-target system to a multi-target system. The multiple-target framework has accelerated

 DOI: 10.1201/9781003672760-6

with advancements in computational chemistry, biochemistry, structural biology, molecular biology, and artificial intelligence (AI). Computer-aided drug design (CADD) and in silico methodologies have emerged as important technologies capable of reducing drug discovery and development costs (Lin et al., 2020; Cichońska et al., 2024). Conditions related to metabolic disorders and CVDs exhibit a great interconnection at the molecular level, such as dysregulation of glucose metabolism, lipid homeostasis, dysfunction of adipose tissue, and activation of the sympathetic nervous system, which leads to problems like obesity, insulin resistance, and hypertension (Zhang et al., 2024). So, diseases related to the cardiac system and metabolism are ideal models for the study of advancement in silico drug design methodologies.

6.2 WHAT IS "IN SILICO DRUG DESIGNING?"

The term "In Silico" is derived from the Latin phrases "in vivo (in life)" and "in vitro (in glass)". It refers to experiments performed on silicon chips, that is, computer processors. Traditionally, scientists would physically test thousands of random compounds to see if one worked. This is slow, expensive, and often fails. In the in silico approach, scientists use digital models to scan the haystack instantly. They can filter through millions of virtual chemical structures to predict which ones might work, leaving only the most promising options to be tested. In simple terms, in silico drug design is the process of testing and designing potential medicines on a computer rather than in a physical laboratory (Wadood et al., 2013) (Figure 6.1).

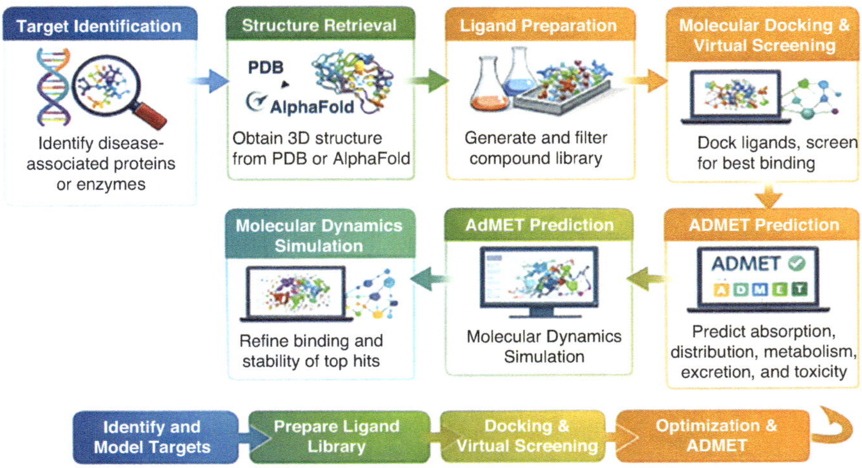

Figure 6.1 General workflow of in silico drug design for metabolic and cardiovascular diseases.

Table 6.1 Generalized working mechanism of in silico drug designing

Step 01	Target identification	First identification of a target in the body that causes a disease, like a specific protein or enzyme. Then create a 3D digital model of this target.	Wadood et al. (2013)
Step 02	Molecular docking	Make digital models of potential drugs for the target. Then take digital models of potential drugs and try to dock them into the target. It's like trying thousands of keys in a lock to see which one turns it perfectly.	Morris & Lim-Wilby, 2008
Step 03	Virtual screening	Instead of testing 1,000 chemicals in a lab, scan a library of 1,000,000 digital chemicals in a database to see which ones are worth making.	Reddy et al. (2007)

Computers can assist in lieu of experimenting with thousands of chemicals by:

a. Identifying an appropriate biological target (a protein/enzyme/receptor that is associated with a particular disease).
b. Creating an accurate model of the target (the three-dimensional shape of the target obtained from either PDB, AlphaFold or homology modeling).
c. Obtaining or designing molecules (drug-like chemical compounds from compound libraries).
d. Determining which molecules will bind to the target through molecular docking (simulated "key–lock" fitting between the molecule and the target).
e. Virtual screening millions of chemicals to identify potential drug candidates.
f. Using molecular dynamics simulation to further refine the compounds identified as potentially useful based on the virtual screening process.
g. Prediction of ADMET/toxicity properties of the proposed drugs (absorption, distribution, metabolism, excretion, toxicity) (Tables 6.1–6.3).

6.3 OVERVIEW OF METABOLIC AND CARDIOVASCULAR DISORDERS

Metabolic and cardiovascular disorders are diseases that are chronically and non-communicably linked closely together, and all of them are majorly driven by the same risk factors, namely, central obesity, insulin resistance, dyslipidemia, hypertension, and chronic low-grade inflammation

Table 6.2 Classical in silico methods in drug discovery

Structure-based drug design (SBDD)	It leverages the three-dimensional structures of target macromolecules, such as proteins, to design ligands that complement their binding sites rationally. This approach, enabled by X-ray crystallography and NMR, has revolutionized drug discovery by transitioning from random screening to precise modeling of receptor–ligand interactions.	Blundell (1996)
Ligand-based drug design (LBDD)	It exploits the structural and physicochemical properties of known ligands that bind to a target protein, deriving predictive models from structural activity relationship data without needing the target's three-dimensional structure.	Acharya et al. (2011)
Quantitative structure–activity relationship (QSAR)	It establishes mathematical correlations between a compound's chemical structure, quantified via molecular descriptors, and its biological activity or physicochemical properties. This approach assumes that structurally similar molecules exhibit analogous activities, enabling predictive models for potency, toxicity, or ADMET profiles in drug discovery.	Peter et al. (2019)

Table 6.3 Advanced computational methods and technologies in drug discovery

AlphaFold	AlphaFold's breakthrough in protein structure prediction promises to streamline drug discovery by providing rapid, accurate 3D models from amino acid sequences alone. This addresses longstanding bottlenecks where experimental methods like X-ray crystallography often took years and failed for many targets. AlphaFold's CASP14 success is important in enabling faster target validation and ligand design.	Mullard (2021)
Graph neural networks (GNNs)	GNNs have become pivotal in AI-aided drug discovery. GNNs excel in processing graph-structured data like molecular representations, enabling message passing for feature aggregation. Advances include geometric GNNs for 3D conformations, interpretable variants for mechanistic insights, and scalable architectures handling large datasets, alongside uncertainty quantification to flag unreliable predictions.	Zhang et al. (2025)

(Madhu & Sarkar, 2016; Sarkar et al., 2021, 2022; Mohamed et al., 2023; Huang, 2009; Madhual et al., 2023; Modi et al., 2024; Kaur et al., 2024; Ranganathan et al., 2024; Swarup et al., 2025; Hamooya et al., 2025). The core metabolic disorders, which are type 2 diabetes mellitus, obesity, hyperlipidemia, and the metabolic syndrome, can be characterized by the disruption of glucose and lipid homeostasis, visceral adiposity, and atherogenic lipid profiles (Bonora, 2006; Tkáč, 2005; Bjornstad & Eckel, 2018; Ritika Patial et al., 2024). All these factors multiply the possibility of an individual suffering from atherosclerotic CVD and heart failure (Rana et al., 2007; Alegría Ezquerra et al., 2008). On the contrary, conditions affecting the heart and blood vessels, including hypertension and atherosclerosis, have relatively frequently developed as the macrovascular complications of metabolic dysfunction that has been going on for a long time, thus making cardiometabolic multimorbidity the leading global health burden recognized today (Kirk & Klein, 2009; Dong et al., 2025). The point at which these disorders collide in their pathophysiological aspect is where a limited number of molecular and cellular targets are available (Fredman et al., 2014; Welty et al., 2016). Among these are the enzymes that control the metabolism of glucose and lipids, nuclear and membrane receptors (e.g., insulin, angiotensin, and incretin receptors), transporters such as sodium–glucose cotransporters and lipid transport proteins, and cellular signaling pathways of oxidative stress, endoplasmic reticulum stress, and inflammation that are pro-atherogenic or pro-hypertrophic (Schulman, 2010; Jin et al., 2025; Ugariogu et al., 2025). The fact that there is a common mechanistic landscape among the three conditions gives a reason for in silico drug design, which can be employed to discover and refine small molecules or biologics that act on enzymes, receptors, transporters, and signaling networks with the dual purpose of correcting metabolic disturbances and, at the same time, providing cardiovascular protection (Fredman et al., 2014; Rochlani et al., 2017; Ugariogu et al., 2025).

6.4 COMPUTATIONAL APPROACHES FOR DRUG DEVELOPMENT

The integration of computational methods into drug discovery has reached a point where it is almost impossible to do without them because they help to design and optimize the molecules that are taking the lead as candidates, all this before the costly experimental testing is done (Ferreira et al., 2015; Meng et al., 2011; Adelusi et al., 2022). In SBDD, docking of the molecule, molecular dynamic simulations, and structure-based pharmacophore modeling are guided by the high-resolution structures or homological models of the main metabolic or cardiovascular targets of the drug (Ferreira et al., 2015; Meng et al., 2011; Opo et al., 2021; Oduselu et al., 2023).

This approach then predicts the binding modes, refines the ligand–target interactions, and examines the receptor's flexibility at the atomic level (Zare et al., 2024; Luo et al., 2021). On the other hand, LBDD coincides with SBDD whenever structural data is scarce by resorting to QSAR and 3D-QSAR techniques like CoMFA and CoMSIA to correlate molecular descriptors with bioactivity and to create contour maps that show the steric, electrostatic, hydrophobic, and hydrogen-bonding features essential for potency and selectivity (Acharya et al., 2011; Srivastava et al., 2008; Ferreira da Cunha et al., 2009; Gunda et al., 2015). These predictive models form the basis of virtual screening campaigns, which involve the sifting through large libraries of compounds using structure-based docking workflows or ligand-based similarity and pharmacophore searches to rapidly rank hits with the most favorable interaction patterns against cardiometabolic targets (Opo et al., 2021; Oduselu et al., 2023; Luo et al., 2021). At last, in silico ADME and toxicity prediction tools combine cheminformatics and machine learning to predict the ADMET properties, which include absorption, distribution, metabolism, excretion, and toxicity, respectively (Wang et al., 2015; Han et al., 2025; Xiong et al., 2021). As a result of these predictions, compounds with poor pharmacokinetics or safety issues are eliminated early, thereby increasing the effectiveness and success rate of candidate selection in cardiometabolic drug discovery (Rekha Samal et al., 2024; Zare et al., 2024).

6.5 MOLECULAR TARGETS FOR METABOLIC AND CARDIOVASCULAR DISEASES

Molecular targets implicated in metabolic and CVDs include enzymes, nuclear receptors, and signaling proteins that collectively regulate glucose and lipid homeostasis, vascular tone, and cardiac function (Chalotra et al., 2025; Domingo et al., 2024). In metabolic disorders, dipeptidyl peptidase-4 (DPP-4) degrades incretin hormones and thereby modulates insulin secretion and cardiometabolic inflammation (Gong et al., 2015; Chen et al., 2022; Zhang et al., 2019). Peroxisome proliferator-activated receptor gamma (PPAR-γ) controls the transcription of genes involved in adipogenesis, insulin sensitivity, and lipid metabolism, making both key targets for antidiabetic and anti-obesity therapies (Vázquez-Carrera & Wahli, 2022; Lonardo & Weiskirchen, 2025; Lin et al., 2022). AMP-activated protein kinase (AMPK) functions as a cellular energy sensor that coordinates glucose uptake, fatty acid oxidation, and cholesterol synthesis. Among its downstream effects is the inhibition of HMG-CoA reductase, the rate-limiting enzyme in cholesterol biosynthesis targeted by statins to reduce low-density lipoprotein levels and atherosclerotic risk (Srivastava et al., 2012; Badmus et al., 2022). On the cardiovascular side, angiotensin-converting enzyme (ACE) generates the vasoconstrictor angiotensin II, whereas β-adrenergic receptors (β-ARs) and endothelin receptors orchestrate sympathetic and endothelin-mediated control of heart rate, contractility, and vascular tone, together forming central nodes in hypertension and heart failure pharmacotherapy (Conti et al., 2013; Vaniotis et al., 2013; Moens et al., 2010). Endothelial nitric oxide synthase (eNOS), the predominant NOS isoform in the vasculature, produces nitric oxide to maintain endothelial function and blood pressure homeostasis. Its dysfunction is closely linked to hypertension, atherosclerosis, myocardial infarction, and stroke, highlighting eNOS and its regulatory network as crucial vascular-protective targets (Tran et al., 2022; Li et al., 2025).

6.6 STRUCTURE RETRIEVAL AND PROTEIN PREPARATION

The process of structure retrieval and protein preparation is the first and foremost step in the in silico drug design process for metabolic and cardiovascular targets. It ensures that the three-dimensional model used for docking and simulation is precise, complete, and biophysically reasonable (Chang et al., 2022; Meng et al., 2011). The experimentally determined structures are usually acquired from the Protein Data Bank (PDB), while UniProt offers well-organized protein sequences and functional annotations that can be linked to the available structures (Martin, 2005). For proteins without experimental coordinates, excellent homology models can be made using automated servers like SWISS-MODEL and accessed from its model repository (Bordoli & Schwede, 2012; Kiefer et al., 2009). Protein preparation carries out the cleaning of the structure (removal of crystallographic artifacts, non-essential ligands, and bulk water), atom and hydrogen addition, protonation state assignment, and the conducting of energy minimization or limited refinement to relieve steric clashes and optimize local geometry for reliable docking calculations once a suitable template structure or model is chosen (Verburgt & Kihara, 2022; Osoro & Onsinyo, 2025). Furthermore, binding site identification and validation are achieved either by using co-crystallized ligands in protein structures or by employing computational pocket-detection and druggability tools (such as Sitemap and related algorithms) that analyze protein surface shape, chemical and physical properties, and solvent exposure to identify potential cavities and determine their suitability as small molecule targets (Guo et al., 2015; Liao et al., 2022; Fatima et al., 2025).

6.7 LIGAND PREPARATION AND LIBRARY DEVELOPMENT

Ligand preparation and library development provide the chemical foundation for in silico drug design by assembling and standardizing sets of small molecules suitable for virtual screening against metabolic and cardiovascular targets (Shoichet, 2004; Oprea, 2002; Singh et al., 2020). Natural compound libraries derived from medicinal plants, marine organisms, and microbial sources offer structurally diverse, stereochemically rich scaffolds for cardiometabolic lead discovery (Santana et al., 2021; Sokouti, 2023). Synthetic and combinatorial libraries, including ultra-large "on-demand" collections, enable systematic exploration of drug-like chemical space and rapid optimization of structure–activity relationships (Grotsch et al., 2024; Zhou et al., 2024). Before screening, all ligands must undergo rigorous data curation and format conversion, typically involving the removal of duplicates, salts, and inconsistent records, normalization of valence and stereochemistry, and transformation between SMILES, SDF, and other 2D/3D representations using cheminformatics toolkits. This ensures that clean, non-redundant structures can be reliably protonated, energy-minimized, and conformationally sampled (Barreto Gomes et al., 2023; Prabhakaran et al., 2023). To enhance the oral bioavailability of candidates and decrease late-stage attrition, libraries are often filtered using drug-likeness criteria such as Lipinski's rule of five and related property-based filters. These criteria limit molecular weight, lipophilicity, and hydrogen-bond counts, thereby prioritizing compounds with acceptable permeability and pharmacokinetic profiles for cardiometabolic indications (Ani et al., 2018; Nhlapho et al., 2024).

6.8 CASE STUDIES IN METABOLIC DISEASE DRUG DESIGN

In silico techniques are enhancing the ability to identify new therapeutics as well as rationally selecting targets for metabolic disorders. They involve virtually screening thousands of compounds and assessing the pharmacokinetics and safety of lead compounds early on. A number of case studies illustrate how the use of computer algorithms can help researchers understand the complexities of metabolic diseases and develop drugs to treat them.

6.8.1 DPP-4 inhibitors for type 2 diabetes mellitus

The dipeptidyl peptidase-4 (DPP-4) is a key enzyme in the breakdown of incretin hormones that regulate insulin release and blood sugar levels. Researchers used in silico techniques, including virtual screening and docking, to find novel DPP-4 inhibitors in large collections of natural and synthetic compounds. Once lead compounds were found, researchers evaluated the stability of the resulting complexes between the ligands and enzymes through molecular dynamics simulation and the use of ADMET prediction tools to eliminate compounds with had undesirable pharmacokinetic characteristics. Ultimately, these in silico approaches allowed researchers to find lead compounds that bound more effectively to DPP-4 than previous compounds and demonstrated greater specificity.

6.8.2 PPAR-γ modulators for treating obesity and insulin resistance

PPAR-γ (peroxisome proliferator-activated receptor gamma) is a nuclear hormone receptor that regulates fat cell formation, lipid metabolism, and insulin sensitivity. Researchers have used various in silico drug design strategies to identify novel PPAR-γ modulators that are less likely to cause unwanted side effects than the full agonist drugs currently in use. Ligand-based QSAR and pharmacophore models have been used to determine structural requirements for receptor activation; docking and molecular dynamics simulations have been used to examine ligand-receptor binding modes and the receptor's conformational changes upon ligand binding. Collectively, these in silico studies have led to the identification of partial agonists and dual-acting compounds that exhibit better metabolic efficacy and safety profiles than the current full agonist drugs.

6.8.3 AMPK activators for treating metabolic syndrome

AMPK, a key sensor of cellular energy status, is an attractive target for the treatment of metabolic syndrome. Researchers have screened large collections of natural and synthetic compounds using in silico screening techniques, such as docking-based binding energy evaluations and molecular dynamics simulations, to identify novel AMPK activators. Additionally, researchers have computationally examined the allosteric binding sites of AMPK, providing new avenues for selectively activating this kinase. Furthermore, integrating ADMET predictions has allowed researchers to prioritize compounds with favorable oral bioavailability and metabolic stability.

6.8.4 HMG-CoA reductase inhibitors for hypercholesterolemia

HMG-CoA reductase, the rate-limiting enzyme in cholesterol biosynthesis, has been the subject of extensive in silico research to improve lipid-lowering therapies. Both structure-based drug design and virtual screening have been applied to discover novel statin-like compounds and non-statins. Molecular docking and binding free-energy calculations have optimized ligand interactions at the active site, while computational toxicity assessments have minimized off-target effects, specifically hepatotoxicity.

6.8.5 Multi-target approaches to treating metabolic syndrome

Because metabolic syndrome is a multifactorial disorder, researchers have begun to apply network pharmacology and systems-level in silico approaches to identify compounds that can modulate multiple metabolic pathways simultaneously. By combining in silico target prediction, pathway analysis, and docking studies, researchers have been able to identify both natural products and synthetic compounds that possess polypharmacological activity. This strategy offers promising approaches to developing holistic treatments that address both insulin resistance and inflammation as well as dyslipidemia.

Collectively, the above examples demonstrate that in silico drug design is an essential component of metabolic disease research because it allows for rapid lead discovery, provides a rational approach to optimizing lead compounds, and reduces the late-stage attrition of potential drug candidates. The integration of structure-based modeling, ligand-based analysis, molecular dynamics simulations, and ADMET prediction has proven to be a very successful combination in identifying safe and effective drug candidates for complex metabolic disorders.

6.9 CASE STUDIES IN CARDIOVASCULAR DISEASE DRUG DESIGN

Utilizing in silico drug design has dramatically increased the discovery and development of cardiovascular therapeutics by providing methods to screen potential drug candidates specifically against certain targets, provide mechanistic information about how drug candidates interact with their targets, and provide information on the potential side effects of drug candidates before they enter preclinical testing.

There are several examples of successful applications of in silico drug design in CVD drug design, as presented below in five separate cases.

6.9.1 Angiotensin-converting enzyme (ACE) inhibitors for hypertension

In addition to its role in regulating blood pressure through the conversion of angiotensin I to angiotensin II, the ACE contains a zinc ion that binds to various substrates. The zinc ion is contained within a cleft that includes several other residues involved in substrate binding. This structure has led to the application of in silico methods such as molecular docking and virtual screening to develop new ACE inhibitors from both natural and synthetic sources. These in silico approaches have also enabled researchers to study the interactions between ACE and its substrates, thereby modifying the chemical structure of ACE inhibitors to improve their potency and specificity. Additionally, molecular dynamics simulations have shown that ACE-inhibitor complexes are stable under physiological conditions.

6.9.2 β-Adrenergic receptor modulators in heart failure

β-ARs play a crucial role in regulating heart rate and contractility and thus represent significant targets for managing patients with heart failure. In silico drug design has greatly contributed to our understanding of the interactions between β-ARs and their ligands at the molecular level, using methods such as homology modeling and docking for the β_1- and β_2-subtypes of β-ARs. Additionally, in silico screening techniques have identified specific β_1-AR modulators that exhibit fewer off-target effects than previously developed agents. Furthermore, molecular dynamics simulations have described the structural changes that occur when β-ARs are activated as agonists and inhibited as antagonists.

6.9.3 Endothelin receptor antagonists for pulmonary and systemic hypertension

Endothelin-1 is a powerful vasoconstrictor that contributes to the pathophysiology of pulmonary arterial hypertension and vascular remodeling. Structure-based methods such as virtual screening and pharmacophore modeling have been used to develop new endothelin receptor antagonists for ETA and ETB receptors. Computational studies have allowed researchers to optimize the selectivity and binding affinity of these ligands for specific endothelin receptors. Early identification of favorable pharmacokinetic and safety properties of these ligands was also possible using ADMET prediction tools.

6.9.4 eNOS Modulators for endothelial dysfunction

eNOS, which produces nitric oxide, is critical for maintaining vascular homeostasis. Small molecule enhancers of eNOS activity or stabilizers of eNOS coupling have been designed using computational methods to restore endothelial function in disease states. Insights into ligand binding to the regulatory domains of eNOS were obtained through docking and molecular dynamics studies. Network-based approaches have also identified indirect modulators of eNOS activity that act through upstream signaling pathways.

6.9.5 hERG Liability screening in cardiovascular drug development

Drug-induced QT prolongation due to blockade of the hERG potassium channel represents a leading cause of cardiovascular-related late-stage drug failures. Therefore, in silico cardiotoxicity screening has evolved as an essential component of cardiovascular drug design. A variety of computational models, including structure-based models, QSAR analysis, and deep learning frameworks, have been developed to predict hERG channel binding and the potential for proarrhythmia. Use of these predictive tools enables early identification and removal of potentially toxic compounds from drug development pipelines, thereby enhancing the safety profiles of candidate cardiovascular drugs.

The case studies above demonstrate the significance of in silico drug design in CVD drug design. Methods such as molecular docking, molecular dynamics simulations, pharmacophore modeling, and toxicity prediction have enhanced the effectiveness, precision, and safety of cardiovascular drug discovery efforts. Since CVDs often involve multiple mechanisms and/or targets, computational approaches will continue to be useful in identifying effective treatments for these complex disorders and will help ensure that therapeutic candidates do not exhibit cardiotoxicity prior to initiation of clinical trials.

6.10 ADMET, TOXICITY, AND COMPUTATIONAL PHARMACOKINETICS

ADMET, toxicity, and computational pharmacokinetics analysis are major preclinical steps in the process of *in silico* drug design, enabling the early prediction of absorption, distribution, metabolism, excretion, and toxicity profiles for metabolic and cardiovascular candidates at the discovery stage (Kar & Leszczynski, 2020; Komura et al., 2023; Knoll et al., 2022). The use of *in silico* ADME methods is based on the application of quantitative structure–property relationships and machine learning, wherein large experimental datasets are used for training, to make direct estimates of various pharmacokinetic endpoints like oral absorption, blood–brain barrier penetration, volume of distribution, metabolic stability, and clearance from the chemical structure (Pires et al., 2015; Swanson et al., 2024; Zhou et al., 2024). There are several web platforms such as SwissADME, pkCSM, and ADMETlab that provide users with integrated workflows that take SMILES or structural files as input and produce panels of predicted physicochemical descriptors, ADME properties, and toxicity flags as output (AL Azzam, 2022; Dulsat et al., 2023; Fu et al., 2024). Thus, profiling large virtual libraries can be performed rapidly and at very low cost (Kar & Leszczynski, 2020; Mahgoub et al., 2022). The predictions are then combined with a rule-based approach for drug-likeness and toxicity filters, allowing for the elimination of non-druglike, promiscuous, or potentially toxic molecules before docking or experimental testing, which directs the resources toward the compounds with favorable pharmacokinetic and safety profiles (Mahgoub et al., 2022; Mou et al., 2025).

Early prediction of cardiotoxicity is of utmost importance in the case of drug discovery for metabolic and CVDs, and therefore the application of specific models such as hERG potassium channel blockade and QT-interval prolongation, including hERG-focused QSAR tools, deep learning frameworks like DeepHIT, and multiscale cardiac risk pipelines is on the rise to detect compounds with proarrhythmic potential and minimize the risks of late-stage failure or market withdrawal due to the associated issues (Chemi et al., 2017; Creanza et al., 2021; Ryu et al., 2020; Schober et al., 2025; Yang et al., 2020).

6.11 NETWORK PHARMACOLOGY AND SYSTEMS BIOLOGY APPROACHES

Network pharmacology and systems biology are in silico, multi-omics, network-based approaches that move beyond "one drug–one target" to tackle complex metabolic and CVDs as network disorders (Li et al., 2023; Wang et al., 2023; Reitz et al., 2023; Yue & Dutta, 2022; Mukherjee et al., 2024).

6.11.1 Conceptual basis and network construction

Network pharmacology treats disease as a perturbation of interacting molecular networks and drugs as network modulators acting on multiple nodes in PPI, regulatory, and metabolic networks (Li et al., 2023; Yue & Dutta, 2022; Mukherjee et al., 2024). Disease genes, drug–target interactions, and pathways from resources such as STRING-like PPIs, KEGG/Reactome pathways, and DrugBank-type pharmacology are integrated to build disease and drug–target networks, enabling identification of key hubs and bottlenecks that drive cardiometabolic pathology (Reitz et al., 2023; Agamah et al., 2022; Yue & Dutta, 2022; Doran et al., 2021; Mukherjee et al., 2024) (Table 6.4 and Figure 6.2).

6.11.2 Applications in metabolic and cardiovascular disease

- Multi-omics network medicine in CVD integrates epigenome, transcriptome, proteome, and metabolome to explain heterogeneity and guide precision therapeutics (Wang et al., 2023; Reitz et al., 2023; Lin et al., 2025; Doran et al., 2021).
- Network-based multi-omics frameworks link lipids, metabolites, and proteins to CVD modules and drug effects (e.g., empagliflozin's time-resolved impact on lipid metabolism) (Anyaegbunam et al., 2025).

Table 6.4 Network pharmacology and systems biology approaches in cardiometabolic drug discovery

Concept	*Role in cardiometabolic disease/drug discovery*	*Citations*
Multi-omics systems biology	Integrates genomics, transcriptomics, proteomics, and metabolomics to model CVD and metabolic disease mechanisms, network dynamics, and feedback	Reitz et al. (2023), Wang et al. (2023), Lin et al. (2025), Yue and Dutta (2022), and Doran et al. (2021)
Network pharmacology/ network medicine	Maps drugs and disease modules on the human interactome to find multi-target interventions, combinations, and repurposing opportunities	Wang et al. (2023), Agamah et al. (2022), Pandey et al. (2025), Li et al. (2023) and Mukherjee et al. (2024)
Reverse/ network-based screening	Uses disease networks to backtrack candidate compounds (e.g., berberine for cardiovascular–kidney–metabolic syndrome)	Wen et al. (2025) and Meng et al. (2026)

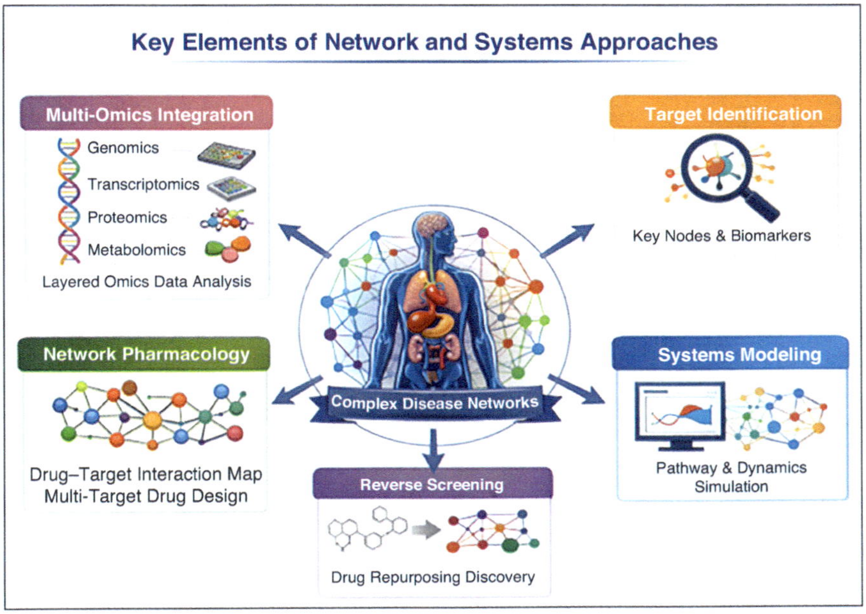

Figure 6.2 Key elements of network and systems approaches.

- Network pharmacology and integrative bioinformatics dissect multi-component natural products and food–medicine homology plants in coronary heart disease and cardiovascular–kidney metabolic syndrome, combining PPI networks, WGCNA, docking, dynamics, and ADMET to reveal multi-target immunometabolic mechanisms (Meng et al., 2026; Wen et al., 2025).

6.11.3 Advantages, challenges, and future directions

These approaches help identify synergistic target sets, predict off-targets and adverse effects, enable drug repurposing, and support personalized therapy by stratifying CVD endotypes via network and multi-omics signatures (Wang et al., 2023; Reitz et al., 2023; Yue & Dutta, 2022; Pandey et al., 2025). Key challenges include incomplete/biased networks, noisy heterogeneous data, and limited experimental/clinical validation; emerging directions emphasize AI/ML-driven integration, dynamic network modeling, and patient-level or single-cell multi-omics for truly personalized cardiometabolic interventions (Agamah et al., 2022; Zhang et al., 2023; Lin et al., 2025; Norzagaray-Valenzuela et al., 2025; Holman et al., 2024; Ganekal et al., 2023; Mukherjee et al., 2024).

6.12 ARTIFICIAL INTELLIGENCE IN DRUG DESIGN

AI is being transformed in the modern era through its ability to analyze data, make predictions, and optimize drug discovery across all stages of the drug development process. Through combining ML, DL, and statistical modeling with classical in silico methods, AI is capable of processing vast amounts of biological and chemical data. As many complex diseases, such as metabolic and CVDs, result from interrelated molecular networks rather than single disease targets, AI offers an ideal solution for these types of diseases (Pandey et al., 2025; Wang et al., 2023; Yue & Dutta, 2022).

During the initial stage of drug discovery, AI uses algorithms to identify and validate drug targets through analysis of multi-omics datasets (i.e., genomics, transcriptomics, proteomics, and metabolomics). This type of integrative approach allows researchers to identify genes associated with disease, regulatory pathways, and molecular biomarkers related to cardiometabolic disease progression (Doran et al., 2021; Reitz et al., 2023; Lin et al., 2025). Additionally, AI-driven network analysis facilitates the identification of key regulatory nodes and common transcriptional regulators connecting both metabolic and cardiovascular phenotypes, which identifies actionable therapeutic targets (Ganekal et al., 2023; Norzagaray-Valenzuela et al., 2025).

Additionally, AI has greatly improved the efficiency of virtual screening and lead discovery processes. Deep learning architectures such as those using graph-based neural networks have been able to learn the complexity of structure–activity relationship between molecules and their activity based upon molecular representation. These architectures are able to predict ligand–target interactions at a level of accuracy that surpasses traditional computational screening methods. Furthermore, when combined with network pharmacology, AI has enabled researchers to conduct a system-wide analysis of the interaction between drugs and their targets and pathways involved in the treatment of complex diseases. Therefore, AI enables researchers to rationally design multi-target therapeutics (Li et al., 2023; Zhang et al., 2023; Pandey et al., 2025). Additionally, generative AI frameworks enable researchers to perform de novo molecular design by suggesting novel chemical structures that possess optimal binding affinity, selectivity, and drug-likeness, allowing researchers to rapidly search through vast chemical space that cannot be screened using traditional screening methods (Mukherjee et al., 2024).

Finally, AI has increased importance in pharmacokinetics and toxicological assessments. Large experimental datasets can be used to train machine learning models that can accurately predict ADMET profiles. Using this information, researchers can eliminate compounds that would be unsafe or exhibit poor bioavailability during the early stages of drug development. Cardiovascular drug discovery applications are increasing the use of AI-supported systems and network pharmacology to assess cardiotoxicity risks and design safer, multi-target therapeutic strategies for cardiometabolic syndromes (Agamah et al., 2022; Meng et al., 2026; Wen et al., 2025). The integration of AI with systems biology and network medicine will provide a robust platform for developing next-generation cardiometabolic drugs with improved efficacy and translational potential.

6.13 CHALLENGES, LIMITATIONS, AND TRANSLATIONAL CONSIDERATIONS

While AI has shown considerable potential for drug design, numerous barriers and limitations hinder successful clinical translation. Among the primary concerns are issues with data quality and bias. AI systems rely heavily upon large amounts of high-quality data that have been appropriately annotated. When data is incomplete, biased, or "noisy," it will significantly impact AI system performance in terms of prediction accuracy and the ability of those systems to generalize to different populations of patients.

The second concern is the lack of interpretability in many deep learning systems. As AI systems demonstrate high predictive capabilities, they are often considered "black boxes." As such, it is difficult to understand the biological basis for the predictions these systems make. Therefore, developing frameworks that explain AI-based decisions is necessary to improve transparency and confidence among both experimentalists and regulatory bodies.

In addition to interpretability issues, translational experimental and clinical validation of computational predictions is a critical bottleneck in identifying successful therapeutics. Prior to proceeding to human trials, predictions generated using computational models must be validated using in vitro and in vivo experiments as well as clinical trials to demonstrate efficacy and safety. However, translating AI-generated candidate therapeutics into successful treatments also involves addressing differences in the mechanisms of action between preclinical model systems and the complex biology of human subjects.

Finally, ethical and regulatory considerations are also increasingly becoming important factors in the successful clinical transition of AI-assisted drug discovery. Data privacy and reproducibility issues, along with standardizing workflows in AI-assisted drug discovery, require careful management. Regulatory frameworks for AI-assisted drug discovery are currently under development; therefore, collaborative efforts between researchers, clinicians, and regulatory bodies are required to ensure the responsible and compliant use of AI technologies.

6.14 FUTURE PERSPECTIVES

Drug discovery using AI in the future will be based on the development of hybrid, integrated drug discovery models combining AI with physics-based modeling, experimental validation, and system-level analysis (systems biology).

Advances in explainable AI are anticipated to provide better interpretation of AI models, which would allow for better insight into biological mechanisms and ultimately better decision-making during lead optimization.

Another area of promise is personalized medicine. Combining patient-specific omics data with AI-based modeling could enable the creation of therapeutics tailored to a patient's unique genetics, metabolism, and environment in cardiometabolic disease. Personalized medicine has the potential to increase treatment success rates while reducing adverse events.

As well as the expanded use of high-quality biomedical datasets, the application of Federated Learning technology will enable the development of AI models collaboratively without compromising user data security. As AI technologies continue to evolve, they are expected to seamlessly integrate with network pharmacology, systems biology, and Experimental Platforms to significantly accelerate the identification of new safe and effective treatments for metabolic and CVDs.

6.15 CONCLUSION

The increasing global prevalence of metabolic and CVDs emphasizes the urgency for new and effective methods to discover drugs that can be used to treat the many causes and consequences of these conditions. Because traditional pharmacologic approaches targeting one biological molecule have had limited success in treating multifactorial disorders, there now needs to be a change from single-target to multi-target and system-level drug development paradigms.

As part of this evolution, in silico drug design has become a critical component of cardiometabolic drug discovery today. The use of in silico drug design integrates established computational approaches (e.g., docking) with novel methodologies, including network pharmacology, systems biology, and AI, which facilitate the rational identification of drug targets, rapid virtual screening of compounds, optimization of leads, and early assessment of pharmacokinetic properties and potential toxicities. Overall, these capabilities make it possible to develop drugs much faster and cheaper than was previously possible.

Additionally, the increased use of multi-omics integration, network-based analyses, and AI-based models has provided a better understanding of disease mechanisms and how they interact within the body at both the molecular and systems levels, thereby enabling the development of drugs that act on multiple targets, thereby improving efficacy and reducing adverse effects. Furthermore, the overlap of many metabolic and cardiovascular pathways offers a clear rationale for developing computer-assisted, integrated cardiometabolic treatments.

While significant challenges exist currently regarding data quality, model interpretability, and translational validation, continued advances in explainable AI, the availability of high-quality omic datasets, and the emergence of hybrid computational-experimental platforms will enhance the translation of in silico discoveries into clinically relevant discoveries. Ultimately, as in silico drug design continues to evolve, it will likely assume an increasingly important role in the development of future generations of precision medicines for metabolic and CVDs and will contribute to improved global health care and management.

REFERENCES

Acharya, C., Coop, A., Polli, J.E., & Mackerell, A.D., Jr (2011). Recent advances in ligand-based drug design: relevance and utility of the conformationally sampled pharmacophore approach. *Current Computer-Aided Drug Design*, 7(1), 10–22. https://doi.org/10.2174/157340911793743547

Acharya, C.K., Das, B., Madhu, N.R., Sau, S., Manna De, M., & Sarkar, B. (2023). A Comprehensive Pharmacological Appraisal of Indian Traditional Medicinal Plants with Anti-diabetic Potential. In: *Advances in Diabetes Research and Management*, pp. 163–193. Springer Nature Singapore Pte Ltd. https://doi.org/10.1007/978-981-19-0027-3_8

Adelusi, T.I., Oyedele, A.Q.K., Boyenle, I.D., Ogunlana, A.T., Adeyemi, R.O., Ukachi, C.D., Idris, M.O., Olaoba, O.T., Adedotun, I.O., Kolawole, O.E., Xiaoxing, Y., & Abdul-Hammed, M. (2022). Molecular modeling in drug discovery. *Informatics in Medicine Unlocked, 29*, 100880. https://doi.org/10.1016/j.imu.2022.100880

Agamah, F., Bayjanov, J., Niehues, A., Njoku, K., Skelton, M., Mazandu, G., Ederveen, T., Mulder, N., Chimusa, E., & Hoen, P. (2022). Computational approaches for network-based integrative multi-omics analysis. *Frontiers in Molecular Biosciences*, 9. https://doi.org/10.3389/fmolb.2022.967205

AL Azzam, K. (2022). SwissADME and pkCSM webservers predictors: an integrated online platform for accurate and comprehensive predictions for in silico ADME/T properties of artemisinin and its derivatives. *Kompleksnoe Ispolzovanie Mineralnogo Syra = Complex Use of Mineral Resources, 325*(2), 14–21. https://doi.org/10.31643/2023/6445.13

Alegría Ezquerra, E., Castellano Vázquez, J.M., & Alegría Barrero, A. (2008). Obesity, metabolic syndrome and diabetes: cardiovascular implications and therapy. *Revista Española de Cardiología* (English Edition). https://doi.org/10.1016/S1885-5857(08)60212-1

Ani, R., Manohar, R., Anil, G., & Deepa, O.S. (2018). Virtual screening of drug likeness using tree based ensemble classifier. *Biomed Pharmacol. J., 11*(3), 1513–1519.

Anyaegbunam, U., Vagiona, A., Cate, V., Bauer, K., Schmidlin, T., Distler, U., Tenzer, S., Araldi, E., Bindila, L., Wild, P., & Andrade, M. (2025). A map of the lipid–metabolite–protein network to aid multi-omics integration. *Biomolecules*, 15. https://doi.org/10.3390/biom15040484

Badmus, O.O., Hillhouse, S.A., Anderson, C.D., Hinds, T.D., & Stec, D.E. (2022). Molecular mechanisms of metabolic associated fatty liver disease (MAFLD): functional analysis of lipid metabolism pathways. *Clinical Science, 136*(18), 1347–1366. https://doi.org/10.1042/cs20220572

Barreto Gomes, D.E., Galentino, K., Sisquellas, M., Monari, L., Bouysset, C., & Cecchini, M. (2023). ChemFlow—From 2D chemical libraries to protein–ligand binding free energies. *Journal of Chemical Information and Modeling, 63*(2), 407–411. https://doi.org/10.1021/acs.jcim.2c00919

Biswas, G., Madhu, N.R., Sarkar, B., Paul, S., Erfani, H., & Alam, Q. (2024). Rare Genetic Disorders: Unraveling the Pathophysiology, Gene Mutations, and Therapeutic Advances in Fabry Disease and Marfan Syndrome. In: Umair, M., Rafeeq, M., Alam, Q. (eds) *Rare Genetic Disorders: Advancements in Diagnosis and Treatment*, pp. 199–219. Springer, *Singapore*. https://doi.org/10.1007/978-981-99-9323-9_7

Biswas, T., Behera, B.K., & Madhu, N.R. (2023). Technology in the Management of Type 1 and Type 2 Diabetes Mellitus: Recent Status and Future Prospects. In *Advances in Diabetes Research and Management*, pp. 111–136, 26 pages, Springer Nature Singapore Pte Ltd. https://doi.org/10.1007/978-981-19-0027-3_6

Bjornstad, P., & Eckel, R.H. (2018). Pathogenesis of lipid disorders in insulin resistance: a brief review. *Current Diabetes Reports*, *18*(12), 127. https://doi.org/10.1007/s11892-018-1101-6

Blundell, T.L. (1996). Structure-based drug design. *Nature*, *384*(6604), 23.

Bonora, E. (2006). The metabolic syndrome and cardiovascular disease. *Annals of Medicine*, *38*(1), 64–80. https://doi.org/10.1080/07853890500401234

Bordoli, L., & Schwede, T. (2012). Automated protein structure modeling with SWISS-MODEL workspace and the protein model portal. *Methods in Molecular Biology (Clifton, N.J.)*, *857*, 107–136. https://doi.org/10.1007/978-1-61779-588-6_5

Chalotra, R., Gupta, A., Singh, T.G., & Singh, R. (2025). Metabolic syndrome: risk factors and molecular drug targets. *EXCLI Journal*, *24*, 1193–1223. https://doi.org/10.17179/excli2025-8703

Chang, Y., Hawkins, B.A., Du, J.J., Groundwater, P.W., Hibbs, D.E., & Lai, F. (2022). A guide to in silico drug design. *Pharmaceutics*, *15*(1), 49. https://doi.org/10.3390/pharmaceutics15010049

Chemi, G., Gemma, S., Campiani, G., Brogi, S., Butini, S., & Brindisi, M. (2017) Computational tool for fast in silico evaluation of hERG K+ channel affinity. *Frontiers in Chemistry*, *5*, 7. https://doi.org/10.3389/fchem.2017.00007

Chen, S.Y., Kong, X.Q., Zhang, K.F., Luo, S., Wang, F., & Zhang, J.J. (2022). DPP4 as a potential candidate in cardiovascular disease. *Journal of Inflammation Research*, *15*, 5457–5469. https://doi.org/10.2147/JIR.S380285

Cichońska, A., Ravikumar, B., & Rahman, R. (2024). AI for targeted polypharmacology: the next frontier in drug discovery. *Current Opinion in Structural Biology*, *84*, 102771. https://doi.org/10.1016/j.sbi.2023.102771

Conti, V., Russomanno, G., Corbi, G., Izzo, V., Vecchione, C., & Filippelli, A. (2013). Adrenoreceptors and nitric oxide in the cardiovascular system. *Frontiers in Physiology*, *4*, 321. https://doi.org/10.3389/fphys.2013.00321

Creanza, T.M., Delre, P., Ancona, N., Lentini, G., Saviano, M., & Mangiatordi, G.F. (2021). Structure-based prediction of hERG-related cardiotoxicity: a benchmark study. *Journal of Chemical Information and Modeling*, *61*(9), 4758–4770. https://doi.org/10.1021/acs.jcim.1c00744

Das, S.K., Sen, K., Sanyal, T., Saha, A., & Madhu, N.R. (2024). Flavonoids: A Promising Neuroprotectant and Its Salutary Effects on Age-Related Neurodegenerative Disorders. In: Pathak, S., Banerjee, A. (eds) *Neuroprotective Effects of Phytochemicals in Brain Ageing*, pp. 221–255. *Singapore*. https://doi.org/10.1007/978-981-99-7269-2_11

Domingo, E., Marques, P., Francisco, V., Piqueras, L., & Sanz, M.-J. (2024). Targeting systemic inflammation in metabolic disorders: a therapeutic candidate for the prevention of cardiovascular diseases? *Pharmacological Research*, *200*, 107058. https://doi.org/10.1016/j.phrs.2024.107058

Dong, H., Chang, L., Tian, T., Shi, R., Yu, K., Wang, C., Liu, Z., Jin, Q., Wang, J., He, T., Chen, H., Shao, X., & Deng, Y. (2025). Mechanism-guided pharmacotherapy for cardiometabolic multimorbidity: from pathophysiology to phenotype-prioritized treatment. *Frontiers in Endocrinology*, *16*, 1724965. https://doi.org/10.3389/fendo.2025.1724965

Dorairaj, P., Jeemon, P., & Roy, A. (2016). Cardiovascular diseases in India: current epidemiology and future directions. *Circulation*, *133*, 1605–1620. https://doi.org/10.1161/CIRCULATIONAHA.114.008729

Doran, S., Arif, M., Lam, S., Bayraktar, A., Turkez, H., Uhlén, M., Borén, J., & Mardinoğlu, A. (2021). Multi-omics approaches for revealing the complexity of cardiovascular disease. *Briefings in Bioinformatics*, *22*. https://doi.org/10.1093/bib/bbab061

Dulsat, J., López-Nieto, B., Estrada-Tejedor, R., & Borrell, J.I. (2023). Evaluation of free online ADMET tools for academic or small biotech environments. *Molecules (Basel, Switzerland)*, *28*(2), 776. https://doi.org/10.3390/molecules28020776

Fatima, I., Safhi, A.Y., Alsalhi, A., Hanbashi, A., Jali, A., Alsabei, H.M., Alshabrmi, F.M., & Sabei, F.Y. (2025). Identification and validation of key covalent inhibitors targeting lung cancer proteins through integrated in silico analysis. *Biochemical and Biophysical Research Communications*, *792*, 152973. https://doi.org/10.1016/j.bbrc.2025.152973

Ferreira, L.G., Dos Santos, R.N., Oliva, G., & Andricopulo, A.D. (2015). Molecular docking and structure-based drug design strategies. *Molecules (Basel, Switzerland)*, *20*(7), 13384–13421. https://doi.org/10.3390/molecules200713384

Ferreira da Cunha, E.F., Sippl, W., Ramalho, T.D.C., Antunes, O.A.C., de Alencastro, R.B., & Albuquerque, M.G. (2009). 3D-QSAR CoMFA/CoMSIA models based on theoretical active conformers of HOE/BAY-793 analogs derived from HIV-1 protease inhibitor complexes. *European Journal of Medicinal Chemistry*, *44*(11), 4344–4352. https://doi.org/10.1016/j.ejmech.2009.05.016

Fredman, G., Ozcan, L., & Tabas, I. (2014). Common therapeutic targets in cardiometabolic disease. *Science Translational Medicine*, *6*(239). https://doi.org/10.1126/scitranslmed.3008908

Fu, L., Shi, S., Yi, J., Wang, N., He, Y., Wu, Z., Peng, J., Deng, Y., Wang, W., Wu, C., Lyu, A., Zeng, X., Zhao, W., Hou, T., & Cao, D. (2024). ADMETlab 3.0: an updated comprehensive online ADMET prediction platform enhanced with broader coverage, improved performance, API functionality and decision support. *Nucleic Acids Research*, *52*(W1), W422–W431. https://doi.org/10.1093/nar/gkae236

Ganekal, P., Vastrad, B., Vastrad, C., & Kotrashetti, S. (2023). Identification of biomarkers, pathways, and potential therapeutic targets for heart failure using next-generation sequencing data and bioinformatics analysis. *Therapeutic Advances in Cardiovascular Disease*, *17*. https://doi.org/10.1177/17539447231168471

Gong, Q., Rajagopalan, S., & Zhong, J. (2015). Dpp4 inhibition as a therapeutic strategy in cardiometabolic disease: incretin-dependent and -independent function. *International Journal of Cardiology*, *197*, 170–179. https://doi.org/10.1016/j.ijcard.2015.06.076

Grotsch, K., Sadybekov, A.V., Hiller, S., Zaidi, S., Eremin, D., Le, A., Liu, Y., Smith, E.C., Illiopoulis-Tsoutsouvas, C., Thomas, J., Aggarwal, S., Pickett, J.E., Reyes, C., Picazo, E., Roth, B.L., Makriyannis, A., Katritch, V., & Fokin, V.V. (2024). Virtual screening of a chemically diverse "superscaffold" library enables ligand discovery for a key GPCR target. *ACS Chemical Biology*, *19*(4), 866–874. https://doi.org/10.1021/acschembio.3c00602

Gunda, S.K., Mutya, S.S., Durgam, S., Adepu, S., & Shaik, M. (2015). Application of 3D QSAR CoMFA/CoMSIA and in silico docking studies on potent inhibitors of interleukin-2 inducible T-cell kinase (ITK). *International Journal of Pharmaceutical Sciences Review and Research*, *30*(1), 28–34.

Guo, Z., Li, B., Cheng, L.-T., Zhou, S., McCammon, J.A., & Che, J. (2015). Identification of protein–ligand binding sites by the level-set variational implicit-solvent approach. *Journal of Chemical Theory and Computation, 11*(2), 753–765. https://doi.org/10.1021/ct500867u

Hamooya, B.M., Siame, L., Muchaili, L., Masenga, S.K., & Kirabo, A. (2025). Metabolic syndrome: epidemiology, mechanisms, and current therapeutic approaches. *Frontiers in Nutrition, 12*, 1661603. https://doi.org/10.3389/fnut.2025.1661603

Han, H., Shaker, B., Lee, J.H., Choi, S., Yoon, S., Singh, M., Basith, S., Cui, M., Ahn, S., An, J., Kang, S., Yeom, M.S., & Choi, S. (2025). Employing automated machine learning (AutoML) methods to facilitate the in silico ADMET properties prediction. *Journal of Chemical Information and Modeling, 65*(7), 3215–3225. https://doi.org/10.1021/acs.jcim.4c02122

Holman, A., Tran, S., Destici, E., Farah, E., Li, T., Nelson, A., Engler, A., & Chi, N. (2024). Single-cell multi-modal integrative analyses highlight functional dynamic gene regulatory networks directing human cardiac development. *Cell Genomics, 4*. https://doi.org/10.1016/j.xgen.2024.100680

Huang, P.L. (2009). A comprehensive definition for metabolic syndrome. *Disease Models & Mechanisms, 2*(5–6), 231–237. https://doi.org/10.1242/dmm.001180

Jin, P., Duan, X., Huang, Z., Dong, Y., Zhu, J., Guo, H., Tian, H., Zou, C.-G., & Xie, K. (2025). Nuclear receptors in health and disease: signaling pathways, biological functions and pharmaceutical interventions. *Signal Transduction and Targeted Therapy, 10*(1). https://doi.org/10.1038/s41392-025-02270-3

Kar, S., & Leszczynski, J. (2020). Open access in silico tools to predict the ADMET profiling of drug candidates. *Expert Opinion on Drug Discovery, 15*(12), 1473–1487. https://doi.org/10.1080/17460441.2020.1798926

Kaur, S., Bansal, K., & Kumar, Y. (2024). Machine learning-based prediction system for risk assessment of hypertension using symptoms investigations. *International Journal of Experimental Research and Review, 46*, 139–149. https://doi.org/10.52756/ijerr.2024.v46.011

Kiefer, F., Arnold, K., Künzli, M., Bordoli, L., & Schwede, T. (2009). The SWISS-MODEL repository and associated resources. *Nucleic Acids Research, 37*(suppl_1), D387–D392. https://doi.org/10.1093/nar/gkn750

Kirk, E.P., & Klein, S. (2009). Pathogenesis and pathophysiology of the cardiometabolic syndrome. *Journal of Clinical Hypertension (Greenwich, Conn.), 11*(12), 761–765. https://doi.org/10.1111/j.1559-4572.2009.00054.x

Knoll, K.E., van der Walt, M.M., & Loots, D.T. (2022). *In silico* drug discovery strategies identified ADMET properties of decoquinate RMB041 and its potential drug targets against Mycobacterium tuberculosis. *Microbiology Spectrum, 10*(2), e0231521. https://doi.org/10.1128/spectrum.02315-21

Komura, H., Watanabe, R., & Mizuguchi, K. (2023). The trends and future prospective of in silico models from the viewpoint of ADME evaluation in drug discovery. *Pharmaceutics, 15*(11), 2619. https://doi.org/10.3390/pharmaceutics15112619

Li, B., Li, X., Zeng, Y., Zhou, Z., Zhao, D., Qin, F., Zhang, B., Yao, W., Mao, Y., Zhou, L., Li, K., Zhu, Q., Rong, X., & Guo, J. (2024). Network pharmacology combined with molecular docking and experimental verification to elucidate functional mechanism of Fufang Zhenzhu Tiaozhi against type 2 diabetes mellitus. *Journal of Biomolecular Structure and Dynamics, 42*(24), 13751–13767. https://doi.org/10.1080/07391102.2023.2278082

Li, H., Förstermann, U., Xia, N., Kuntic, M., Münzel, T., & Daiber, A. (2025). Pharmacological targeting of endothelial nitric oxide synthase dysfunction and nitric oxide replacement therapy. *Free Radical Biology and Medicine, 237*, 455–472. https://doi.org/10.1016/j.freeradbiomed.2025.06.009

Li, L., Yang, L., Yang, L., He, C., He, Y., Chen, L., Dong, Q., Zhang, H., Chen, S., & Li, P. (2023). Network pharmacology: a bright guiding light on the way to explore the personalized precise medication of traditional Chinese medicine. *Chinese Medicine, 18*. https://doi.org/10.1186/s13020-023-00853-2

Liao, J., Wang, Q., Wu, F., & Huang, Z. (2022). In silico methods for identification of potential active sites of therapeutic targets. *Molecules (Basel, Switzerland), 27*(20), 7103. https://doi.org/10.3390/molecules27207103

Lin, M., Guo, J., Gu, Z., Tang, W., Tao, H., You, S., Jia, D., Sun, Y., & Jia, P. (2025). Machine learning and multi-omics integration: advancing cardiovascular translational research and clinical practice. *Journal of Translational Medicine*, 23. https://doi.org/10.1186/s12967-025-06425-2

Lin, X., Li, X., & Lin, X. (2020). A review on applications of computational methods in drug screening and design. *Molecules* (*Basel, Switzerland*), *25*(6), 1375. https://doi.org/10.3390/molecules25061375

Lin, Y., Wang, Y., & Li, P.F. (2022). PPARα: an emerging target of metabolic syndrome, neurodegenerative and cardiovascular diseases. *Frontiers in Endocrinology, 13*, 1074911. https://doi.org/10.3389/fendo.2022.1074911

Lonardo, A., & Weiskirchen, R. (2025). PPARs in molecular pathogenesis and drug treatment of type 2 diabetes-related MASLD. *Exploration of Digestive Diseases, 4*, 100590. https://doi.org/10.37349/edd.2025.100590

Luo, L., Zhong, A., Wang, Q., & Zheng, T. (2021). Structure-based pharmacophore modeling, virtual screening, molecular docking, ADMET, and molecular dynamics (MD) simulation of potential inhibitors of PD-L1 from the library of marine natural products. *Marine Drugs*, *20*(1), 29. https://doi.org/10.3390/md20010029

Madhu, N.R., Biswas, G., Paul, S., Adhikari, S., Sarkar, B., Rafeeq, M.M., & Umair, M. (2024). Challenges and Future Opportunities in Rare Genetic Disorders: A Comprehensive Review. In: Umair, M., Rafeeq, M., Alam, Q. (eds) *Rare Genetic Disorders*, pp. 251–275. Springer, *Singapore*. https://doi.org/10.1007/978-981-99-9323-9_9

Madhu, N.R., & Sarkar, B. (2016). Present status of dietary fat and obesity. Review Article. UGC-Sponsored National Seminar on Food Security and Sustainable Nutrition in India: The Present Scenario. Acharya Prafulla *Chandra College, New Barrackpore*. pp. 120–124. (ISBN: 978–93–5268-180–8).

Madhual, S., Nayak, D., Dalei, S., Padhi, T., & Das, N.R. (2023). Assessment of cardiovascular risk factors in male androgenetic alopecia: a case control study in a tertiary care hospital of western Odisha. *International Journal of Experimental Research and Review*, *36*, 425–432. https://doi.org/10.52756/ijerr.2023.v36.037

Mahgoub, R.E., Atatreh, N., & Ghattas, M.A. (2022). Using Filters in Virtual Screening: A Comprehensive Guide to Minimize Errors and Maximize Efficiency. In: Caballero, J. (ed), *Annual Reports in Medicinal Chemistry, 59*, 99–136. https://doi.org/10.1016/bs.armc.2022.09.002

Martin, A.C.R. (2005). Mapping PDB chains to UniProtKB entries. *Bioinformatics, 21*(23), 4297–4301. https://doi.org/10.1093/bioinformatics/bti694

Meng, J., Li, S., Shao, X., Zhang, C., Li, S., Li, X., Ge, X., Yuan, C., Rong, J., Xiong, P., Shi, H., Chu, Y., & Hu, Y. (2026). Integrating network pharmacology and multidimensional bioinformatics: mechanistic exploration of six food and medicine homology plants in coronary heart disease therapy. *Computational Biology and Chemistry, 120*, 108694. https://doi.org/10.1016/j.compbiolchem.2025.108694

Meng, X.Y., Zhang, H.X., Mezei, M., & Cui, M. (2011). Molecular docking: a powerful approach for structure-based drug discovery. *Current Computer-Aided Drug Design*, *7*(2), 146–157. https://doi.org/10.2174/157340911795677602

Modi, K., Singh, I., & Kumar, Y. (2024). A hybrid transfer learning approach using obesity data for predicting cardiovascular diseases incorporating lifestyle factors. *International Journal of Experimental Research and Review, 46*, 1–18. https://doi.org/10.52756/ijerr.2024.v46.001

Moens, A.L., Yang, R., Watts, V.L., & Barouch, L.A. (2010). Beta 3-adrenoreceptor regulation of nitric oxide in the cardiovascular system. *Journal of Molecular and Cellular Cardiology*, *48*(6), 1088–1095. https://doi.org/10.1016/j.yjmcc.2010.02.011

Mohamed, S.M., Shalaby, M.A., El-Shiekh, R.A., El-Banna, H.A., Emam, S.R., & Bakr, A.F. (2023). Metabolic syndrome: risk factors, diagnosis, pathogenesis, and management with natural approaches. *Food Chemistry Advances, 3*, 100335. https://doi.org/10.1016/j.focha.2023.100335

Morris, G. M., & Lim-Wilby, M. (2008). Molecular Docking. Molecular Modeling of Proteins, 365–382. https://doi.org/10.1007/978-1-59745-177-2_19

Mou, M., Zhang, Y., Qian, Y., Zhou, Z., Liao, Y., Niu, T., Hu, W., Chen, Y., Jiang, R., Zhao, H., Dai, H., Zhang, Y., & Fu, T. (2025). Drug like Filter 1.0: an AI powered filter for collectively measuring the drug-likeness of compounds. *Journal of Pharmaceutical Analysis*, *15*(6), 101298. https://doi.org/10.1016/j.jpha.2025.101298

Mukherjee, A., Abraham, S., Singh, A., Balaji, S., & Mukunthan, K. (2024). From data to cure: a comprehensive exploration of multi-omics data analysis for targeted therapies. *Molecular Biotechnology*, *67*, 1269–1289. https://doi.org/10.1007/s12033-024-01133-6

Mullard, A. (2021). What does AlphaFold mean for drug discovery? *Nature Reviews Drug Discovery, 20*(10), 725–728.

Nhlapho, S., Nyathi, M., Ngwenya, B., Dube, T., Telukdarie, A., Munien, I., Vermeulen, A., & Chude-Okonkwo, U. (2024). Druggability of pharmaceutical compounds using Lipinski rules with machine learning. *Sciences of Pharmacy, 3*(4), 177–192. https://doi.org/10.58920/sciphar0304264

Norzagaray-Valenzuela, C., Valdez-Flores, M., Camberos-Barraza, J., De La Herrán-Arita, A., Osuna-Ramos, J., Magaña-Gómez, J., Angulo-Rojo, C., Guadrón-Llanos, A., Aviña-Padilla, K., & Calderon-Zamora, L. (2025). Shared transcriptional regulators and network rewiring identify therapeutic targets linking type 2 diabetes mellitus and hypertension. *Frontiers in Molecular Biosciences*, *12*. https://doi.org/10.3389/fmolb.2025.1621413

Oduselu, G.O., Afolabi, R., Ademuwagun, I., Vaughan, A., & Adebiyi, E. (2023). Structure-based pharmacophore modeling, virtual screening, and molecular dynamics simulation studies for identification of Plasmodium falciparum 5-aminolevulinate synthase inhibitors. *Frontiers in Medicine*, *9*. https://doi.org/10.3389/fmed.2022.1022429

Opo, F.A.D.M., Rahman, M.M., Ahammad, F., Ahmed, I., Bhuiyan, M.A., & Asiri, A.M. (2021). Structure based pharmacophore modeling, virtual screening, molecular docking and ADMET approaches for identification of natural anti-cancer agents targeting XIAP protein. *Scientific Reports, 11*(1). https://doi.org/10.1038/s41598-021-83626-x

Oprea, T.I. (2002). Virtual screening in lead discovery: a viewpoint. *Molecules: A Journal of Synthetic Chemistry and Natural Product Chemistry*, 7(1), 51–62. https://doi.org/10.3390/70100051

Osoro, O.I., & Onsinyo, J.M. (2025). Effects of structure minimization on the docking score and docking time during computational analysis using the chimera software. *International Journal of Biomedical and Clinical Research, 3*(4), 1–6. https://doi.org/10.59657/2997-6103.brs.25.049

Pandey, A.K., Ghiassian, S.D., & Loscalzo, J. (2025). Network-based precision medicine and systems pharmacology. *British Journal of Pharmacology*. https://doi.org/10.1111/bph.70184

Patial, R., Batta, I., Thakur, M., Chander Sobti, R., & K. Agrawal, D. (2024). Etiology, pathophysiology, and treatment strategies in the prevention and management of metabolic syndrome. *Archives of Internal Medicine Research, 7*(4). https://doi.org/10.26502/aimr.0184

Peter, S.C., Dhanjal, J.K., Malik, V., Radhakrishnan, N., Jayakanthan, M., & Sundar, D. (2019). Quantitative structure-activity relationship (QSAR): modeling approaches to biological applications. In: S. Ranganathan, M. Gribskov, K. Nakai, & C. Schönbach (Eds.), *Encyclopedia of Bioinformatics and Computational Biology: ABC of Bioinformatics* (Vol. 2), pp. 661–676. Academic Press.

Pires, D.E.V., Blundell, T.L., & Ascher, D.B. (2015). pkCSM: predicting small-molecule pharmacokinetic and toxicity properties using graph-based signatures. *Journal of Medicinal Chemistry, 58*(9), 4066–4072. https://doi.org/10.1021/acs.jmedchem.5b00104

Prabhakaran, P., Hebbani, A.V., Menon, S.V., Paital, B., Murmu, S., Kumar, S., Singh, M.K., Sahoo, D.K., & Desai, P.P.D. (2023). Insilico generation of novel ligands for the inhibition of SARS-CoV-2 main protease (3CLpro) using deep learning. *Frontiers in Microbiology, 14*. https://doi.org/10.3389/fmicb.2023.1194794

Rana, J.S., Nieuwdorp, M., Jukema, J.W., & Kastelein, J.J.P. (2007). Cardiovascular metabolic syndrome: an interplay of, obesity, inflammation, diabetes and coronary heart disease. *Diabetes, Obesity and Metabolism, 9*, 218–232. https://doi.org/10.1111/j.1463-1326.2006.00594.x

Ranganathan, L., Rajasundaram, A., & Kumar, S.K. (2024). Demographic and lifestyle factors influencing cardiovascular health among construction workers: a cross-sectional analysis. *International Journal of Experimental Research and Review*, *42*, 312–319. https://doi.org/10.52756/ijerr.2024.v42.027

Reddy, A. S., Pati, S. P., Kumar, P. P., Pradeep, H. N., & Sastry, G. N. (2007). Virtual screening in drug discovery-a computational perspective. Current protein and peptide science, 8(4), 329–351. https://doi.org/10.2174/138920307781369427

Reitz, C., Kuzmanov, U., & Gramolini, A. (2023). Multi-omic analyses and network biology in cardiovascular disease. *Proteomics, 23*. https://doi.org/10.1002/pmic.202200289

Rekha Samal, R., Samal, S., Suryakant Mishra, C., Bihari Jena, A., & K. Duttaroy, A. (2024). Molecular docking and ADMET prediction of modified and derivative forms of 2-deoxy-D-glucose against COVID-19 main protease complex: in silico approach. *Fortune Journal of Health Sciences, 07*(01). https://doi.org/10.26502/fjhs.172

Rochlani, Y., Pothineni, N.V., Kovelamudi, S., & Mehta, J.L. (2017). Metabolic syndrome: pathophysiology, management, and modulation by natural compounds. *Therapeutic Advances in Cardiovascular Disease, 11*(8), 215–225. https://doi.org/10.1177/1753944717711379

Roy, P., Ghosh, D., Sanyal, R., Madhu, N.R., & Dey, A. (2024). The Controversy Surrounding Drugs Against Neurodegenerative Disorders: Benefit or Harm? In: Pathak, S., Banerjee, A. (eds) *Neuroprotective Effects of Phytochemicals in Brain Ageing*, pp. 373–386. Springer, *Singapore*. https://doi.org/10.1007/978-981-99-7269-2_17

Roy, R., Chakraborty, A., Jana, K., Sarkar, B., Biswas, P., & Madhu, N.R. (2023). The Broader Aspects of Treating Diabetes with the Application of Nanobiotechnology. In: *Advances in Diabetes Research and Management*, pp. 137–162. Springer Nature Singapore Pte Ltd. https://doi.org/10.1007/978-981-19-0027-3_7

Ryu, J. Y., Lee, M. Y., Lee, J. H., Lee, B. H., & Oh, K.-S. (2020). DeepHIT: A deep learning framework for prediction of hERG-induced cardiotoxicity. *Bioinformatics, 36*(10), 3049–3055. https://doi.org/10.1093/bioinformatics/btaa075

Sadhu, S., Sarkar, B., Paul, T., Sanyal, T., Mitra, A., & Madhu, N.R. (2025). The potential of stem cells in the treatment of diabetes: an up-to-date review. *Biomedical Research and Therapy, 12*(8), 7680–7696. https://doi.org/10.15419/m7mc9595.

Santana, K., do Nascimento, L.D., Lima e Lima, A., Damasceno, V., Nahum, C., Braga, R.C., & Lameira, J. (2021). Applications of virtual screening in bioprospecting: facts, shifts, and perspectives to explore the chemo-structural diversity of natural products. *Frontiers in Chemistry, 9*. https://doi.org/10.3389/fchem.2021.662688

Sarkar, B., Biswas, P., Acharya, C.K., Jana, S.K., Nahar, N., Ghosh, S., Dasgupta, D., Ghorai, S.K., & Madhu, N.R. (2022). Obesity epidemiology: a serious public health concern in India. *Chettinad Health City Medical Journal, 11*(1), 21–28.

Sarkar, B., Ghorai, S.K., Jana, S.K., Dasgupta, D., Acharya, C.K., Nahar, N., Ghosh, S., & Madhu, N.R. (2021). Overweight and obesity in West Bengal: a serious public health issue. *VEETHIKA-An International Interdisciplinary Research Journal, 7*(4), 9–14. https://doi.org/10.48001/veethika.2021.07.04.002

Sarkar, S., Sadhu, S., Roy, R., Tarafdar, S., Mukherjee, N., Sil, M., Goswami, A., & Madhu, N.R. (2023). Contemporary drifts in diabetes management. *International Journal of Applied Pharmaceutics, 15*(2), 1–9. https://doi.org/10.22159/ijap.2023v15i2.46792

Schober, H., Mazoudji, A., Schuster, M., Digles, D., Paparella, M., Ecker, G.F., & Walter, M. (2025). P18–68 InSilify DrugTox: a computational retrospective analysis of hERG potassium channel inhibition and QT prolongation data from pharmaceutical scientific advice and authorization procedures. *Toxicology Letters, 411*(Suppl.), S216–S217. https://doi.org/10.1016/j.toxlet.2025.07.517

Schulman, I.G. (2010). Nuclear receptors as drug targets for metabolic disease. *Advanced Drug Delivery Reviews, 62*(13), 1307–1315. https://doi.org/10.1016/j.addr.2010.07.002

Shoichet, B.K. (2004). Virtual screening of chemical libraries. *Nature*, *432*(7019), 862–865. https://doi.org/10.1038/nature03197

Singh, N., Chaput, L., & Villoutreix, B.O. (2020). Virtual screening web servers: designing chemical probes and drug candidates in the cyberspace. *Briefings in Bioinformatics*, *22*(2), 1790–1818. https://doi.org/10.1093/bib/bbaa034

Sokouti, B. (2023). A review on *in silico* virtual screening methods in COVID-19 using anticancer drugs and other natural/chemical inhibitors. *Exploration of Targeted Anti-tumor Therapy*, *4*(5), 994–1026. https://doi.org/10.37349/etat.2023.00177

Srivastava, R.A., Pinkosky, S.L., Filippov, S., Hanselman, J.C., Cramer, C.T., & Newton, R.S. (2012). AMP-activated protein kinase: an emerging drug target to regulate imbalances in lipid and carbohydrate metabolism to treat cardio-metabolic diseases. *Journal of Lipid Research*, *53*(12), 2490–2514. https://doi.org/10.1194/jlr.R025882

Srivastava, V., Kumar, A., Mishra, B.N., & Siddiqi, M.I. (2008). CoMFA and CoMSIA 3D-QSAR analysis of DMDP derivatives as anti-cancer agents. *Bioinformation*, *2*(9), 384–391. https://doi.org/10.6026/97320630002384

Sur, T., Das, A., Bashar, S., Tarafdar, S., Sarkar, B., & Madhu, N.R. (2023). Biochemical Assay for Measuring Diabetes Mellitus. In: *Advances in Diabetes Research and Management*, pp. 1–20. Springer Nature Singapore Pte Ltd. https://doi.org/10.1007/978-981-19-0027-3_1

Swanson, K., Walther, P., Leitz, J., Mukherjee, S., Wu, J.C., Shivnaraine, R.V., Zou, J. (2024). ADMET-AI: a machine learning ADMET platform for evaluation of large-scale chemical libraries. *Bioinformatics*, *40*(7), btae416. https://doi.org/10.1093/bioinformatics/btae416

Swarup, S., Ahmed, I., & Grigorova, Y. (2025). Metabolic Syndrome. [Updated 2024 Mar 7]. In: *StatPearls* [Internet]. Treasure Island (FL): StatPearls Publishing; 2025 Jan-. Available from: https://www.ncbi.nlm.nih.gov/books/NBK459248/

Tkáč, I. (2005). Metabolic syndrome in relationship to type 2 diabetes and atherosclerosis. *Diabetes Research and Clinical Practice, 68*(Suppl. 1), S2–S9. https://doi.org/10.1016/j.diabres.2005.03.002

Tran, N., Garcia, T., Aniqa, M., Ali, S., Ally, A., & Nauli, S.M. (2022). Endothelial nitric oxide synthase (eNOS) and the cardiovascular system: in physiology and in disease states. *American Journal of Biomedical Science & Research*, *15*(2), 153–177.

Ugariogu, S.N., Al-Tannak, N.F., & Alenezi, S.S. (2025). A review on recent insights on cardiovascular diseases drug design using natural products and in silico studies. *Kuwait Journal of Science, 52*(4), Article 100438. https://doi.org/10.1016/j.kjs.2025.100438

Vaniotis, G., Glazkova, I., Merlen, C., Smith, C., Villeneuve, L.R., Chatenet, D., Therien, M., Fournier, A., Tadevosyan, A., Trieu, P., Nattel, S., Hébert, T.E., & Allen, B.G. (2013). Regulation of cardiac nitric oxide signaling by nuclear β-adrenergic and endothelin receptors. *Journal of Molecular and Cellular Cardiology*, *62*, 58–68. https://doi.org/10.1016/j.yjmcc.2013.05.003

Vázquez-Carrera, M., & Wahli, W. (2022). PPARs as key mediators in the regulation of metabolism and inflammation. *International Journal of Molecular Sciences*, *23*(9), 5025. https://doi.org/10.3390/ijms23095025

Verburgt, J., & Kihara, D. (2022). Benchmarking of structure refinement methods for protein complex models. *Proteins*, *90*(1), 83–95. https://doi.org/10.1002/prot.26188

Wadood, A., Ahmed, N., Shah, L., Ahmad, A., Hassan, H., & Shams, S.J.O.D.D. (2013). In-silico drug design: an approach which revolutionarised the drug discovery process. *OA Drug Design and Delivery*, *1*(1), 3.

Wang, R.-S., Maron, B.A., & Loscalzo, J. (2023). Multiomics network medicine approaches to precision medicine and therapeutics in cardiovascular diseases. *Arteriosclerosis, Thrombosis, and Vascular Biology, 43*(4), 493–503. https://doi.org/10.1161/atvbaha.122.318731

Wang, Y., Xing, J., Xu, Y., Zhou, N., Peng, J., Xiong, Z., Liu, X., Luo, X., Luo, C., Chen, K., Zheng, M., & Jiang, H. (2015). In silico ADME/T modelling for rational drug design. *Quarterly Reviews of Biophysics, 48*(4), 488–515. https://doi.org/10.1017/s0033583515000190

Welty, F.K., Alfaddagh, A., & Elajami, T.K. (2016). Targeting inflammation in metabolic syndrome. *Translational Research: The Journal of Laboratory and Clinical Medicine*, *167*(1), 257–280. https://doi.org/10.1016/j.trsl.2015.06.017

Wen, S., Fu, A., Liu, F., You, Y., Li, L., Xiao, W., Zhong, H., Hong, X., Zhong, X., Hu, Y., & Liu, Z. (2025). Identifying potential drugs for treating cardiovascular-kidney metabolic syndrome via reverse network pharmacology. *Frontiers in Pharmacology, 16*. https://doi.org/10.3389/fphar.2025.1627236

Xiong, G., Wu, Z., Yi, J., Fu, L., Yang, Z., Hsieh, C., Yin, M., Zeng, X., Wu, C., Lu, A., Chen, X., Hou, T., & Cao, D. (2021). ADMETlab 2.0: An integrated online platform for accurate and comprehensive predictions of ADMET properties. *Nucleic Acids Research, 49*(W1), W5–W14. https://doi.org/10.1093/nar/gkab255

Yang, P.C., DeMarco, K.R., Aghasafari, P., Jeng, M.T., Dawson, J.R.D., Bekker, S., Noskov, S.Y., Yarov-Yarovoy, V., Vorobyov, I., & Clancy, C.E. (2020). A computational pipeline to predict cardiotoxicity: from the atom to the rhythm. *Circulation Research*, *126*(8), 947–964. https://doi.org/10.1161/CIRCRESAHA.119.316404

Yue, R., & Dutta, A. (2022). Computational systems biology in disease modeling and control, review and perspectives. *NPJ Systems Biology and Applications, 8*(1). https://doi.org/10.1038/s41540-022-00247-4

Zare, F., Ataollahi, E., Mardaneh, P., Sakhteman, A., Keshavarz, V., Solhjoo, A., & Emami, L. (2024). A combination of virtual screening, molecular dynamics simulation, MM/PBSA, ADMET, and DFT calculations to identify a potential DPP4 inhibitor. *Scientific Reports, 14*(1). https://doi.org/10.1038/s41598-024-58485-x

Zhang, H., Zhou, X.D., Shapiro, M.D., Lip, G.Y.H., Tilg, H., Valenti, L., Somers, V.K., Byrne, C.D., Targher, G., Yang, W., Viveiros, O., Opio, C.K., Mantzoros, C.S., Ryan, J.D., Kok, K.Y.Y., Jumaev, N.A., Perera, N., Robertson, A.G., Abu-Abeid, A., Misra, A., Wongy, Y.J., Ruiz-Úcar, E., Ospanov, O., Celal Kızılkaya, M., Luo, F., Méndez-Sánchez, N., Zuluaga, M., Lonardo, A., Al Momani, H., Toro-Huamanchumo, C.J., Adams, L., Al-Busafi, S.A., Sharara, A.I., Chan, W.-K., Abbas, S.I., Sookoian, S., Treeprasertsuk, S., Ocama, P., Alswat, K., Kong, A.P.-S., Ataya, K., Lim-Loo, M.C., Oviedo, R.J.,·Szepietowsk, O., Fouad, Y., Zhang, H., Abdelbaki, T.N., Katsouras, C.S., Prasad, A., Thaher, O., Ali, A., Molina, G.A., Sung, K.-C., Chen, Q.-F., Lesmana, C.R.A., & Zheng, M.H. (2024). Global burden of metabolic diseases, 1990–2021. *Metabolism: Clinical and Experimental*, *160*, 155999. https://doi.org/10.1016/j.metabol.2024.155999

Zhang, J., Chen, Q., Zhong, J., Liu, C., Zheng, B., & Gong, Q. (2019). DPP-4 inhibitors as potential candidates for antihypertensive therapy: improving vascular inflammation and assisting the action of traditional antihypertensive drugs. *Frontiers in Immunology*, *10*. https://doi.org/10.3389/fimmu.2019.01050

Zhang, O., Lin, H., Zhang, X., Wang, X., Wu, Z., Ye, Q., Zhao, W., Wang, J., Ying, K., Kang, Y., Hsieh, C.-Y., Hou, T., & Hou, T. (2025). Graph neural networks in modern AI-aided drug discovery. *Chemical Reviews*, *125*(20), 10001–10103.

Zhang, P., Zhang, D., Zhou, W., Wang, L., Wang, B., Zhang, T., & Li, S. (2023). Network pharmacology: towards the artificial intelligence-based precision traditional Chinese medicine. *Briefings in Bioinformatics*, *25*. https://doi.org/10.1093/bib/bbad518

Zheng, H., Fridkin, M., & Youdim, M. (2014). From single target to multitarget/network therapeutics in Alzheimer's therapy. *Pharmaceuticals*, *7*(2), 113–135.

Zhou, G., Rusnac, D.-V., Park, H., Canzani, D., Nguyen, H.M., Stewart, L., Bush, M.F., Nguyen, P.T., Wulff, H., Yarov-Yarovoy, V., Zheng, N., & DiMaio, F. (2024). An artificial intelligence accelerated virtual screening platform for drug discovery. *Nature Communications*, *15*(1). https://doi.org/10.1038/s41467-024-52061-7

Chapter 7

Computational design for neurodegenerative disease targets

Mutthuraj Dasegowda[1], *H. D. Sahana*[2], *and Kanthesh M. Basalingappa*[1]

[1]Division of Molecular Biology, SLS, JSS Academy of Higher Education & Research, Mysuru, Karnataka, India

[2]Department of Biochemistry, Oxford Medical College and Hospital, Bengaluru, Karnataka, India

7.1 INTRODUCTION

Neurodegenerative diseases (NDs) are a heterogeneous category of progressive neurological conditions, occurring because of the progressive deterioration of structure and function of neurons in the central nervous system. The disorders are mainly common among the aging population and are a significant health problem in the world because of their rising prevalence, lengthy disease process, and limited therapeutic plans. Various NDs have not been cured, and most of the medications provide only symptomatic relief and not drug modification, despite the extensive research conducted in this field [1].

7.1.1 Neurodegenerative diseases: An overview

Significant neurodegenerative diseases are Alzheimer's disease (AD), Parkinson's disease (PD), Huntington's disease (HD), and amyotrophic lateral sclerosis (ALS). The disorders do have some common pathological hallmarks even though their clinical presentations and the regions of the brain that they affect vary. These are abnormal protein misfolding and aggregation, mitochondrial dysfunction, oxidative stress, neuroinflammation, synaptic failure, and progressive neuronal loss. Protein homeostasis disruption as the key cause of neurodegeneration is highlighted by the deposition of misfolded proteins like amyloid-b and tau in AD or a-synuclein proteins in PD, mutant huntingtin in HD, and TDP-43 or SOD1 in ALS [2].

NDs do not complicate biological pathways with single-gene or single-target defects: these pathways are complex and interconnected at the molecular level. Maladaptation of cellular functions like autophagy, proteasomal degradation, calcium homeostasis, and axonal transport also contributes to

DOI: 10.1201/9781003672760-7

the increased susceptibility of neurons. Moreover, genetic predisposition, epigenetics, and environment are all factors that play a role in the disease onset and progression, making the therapeutic intervention even more complex [3].

Various obstacles such as late clinical diagnosis, poor knowledge of events leading to the onset of the diseases, and the existence of the BBB, which limits the delivery of drugs to the central nervous system, hamper the development of effective therapies to treat the NDs. In addition, the multifactoriality of these conditions requires the discovery of effective molecular targets that can tune the disease-relevant pathways in the absence of off-target toxicity [4].

The computational design methodologies, in this regard, have come to the fore as an effective resource to uncover disease pathogenesis, discover and test new therapeutic targets, and streamline the drug discovery process. In silico approaches can be used to develop a detailed picture of the biology of NDs and offer a rational pathway to development of therapeutics based on their targets by combining genomic, proteomic, structural and systems-level information. This chapter dwells on the importance of computational design to tackle the obstacles related to ND targets and emphasizes the potential of the computational design to inform the next generation of neurotherapeutic approaches [5].

7.1.2 Rationale for computational design in target discovery

Neurodegenerative disorders are too complex and multifactorial, making it particularly challenging to identify effective therapeutic targets. More classical experimental systems of identifying targets are commonly time-consuming, resource-intensive, and lack the capacity to be able to capture dynamic and systems-level interactions that contribute to the progression of diseases. Since neurodegeneration is a complex and interrelated set of molecular pathways, there is an increasing demand to adopt integrative approaches that would be able to process large volumes of biological data and rank targets of disease interest in a logical way [6,7].

Computational design can be applied to provide a potent framework to overcome such challenges since it allows systematic exploration of genomic, transcriptomic, proteomic, and structural data. In silico methods assist in determining possible targets based on sequence conservation, structural properties, and functional domains as well as network connectivity. Such approaches enable scientists to estimate the druggability of the target, determine the availability of the binding site, and estimate the topicality of the target in the early drug discovery phase to minimize the experimental attrition rate [8].

The other major strength of the computational target discovery method is its capability to deal with the heterogeneity associated with NDs. Complex systems such as disease-related pathways, protein–protein interaction networks, and signaling cascades can be modeled using advanced bioinformatics and systems biology tools and can be used to identify nodes of critical interest that could be suitable and effective therapeutic targets. These network-based methods can be especially useful in NDs, in which simultaneous or sequential control of many targets or pathways could be needed to produce disease-modifying effects [9,10]. Moreover, computational design allows a quick hypothesis generation and testing using virtual screening, molecular docking, and molecular dynamics (MD) simulations. These methods give mechanistic details of the interaction between the targets and ligands at the atomic scale and also aid rational optimization of lead compounds. When used together with machine learning (ML) and AI, computational techniques can also make target prioritization even more effective, detecting hidden patterns and forecasting functional relationships that might not be conspicuous with the traditional case of experimental analysis. Computational design is important in recent target discovery of NDs as it enhances efficacy, lowers the cost, and enhances the comprehension of disease biology. Its combination with experimental validation strategies provides a strong and translationally applicable method to the discovery of new and clinically exploitable targets, which will lead to the creation of effective therapeutics against NDs [11].

7.2 MOLECULAR MECHANISMS UNDERLYING NEURODEGENERATION

7.2.1 Protein misfolding and aggregation

Protein misfolding and aggregation can be considered key pathological processes in many NDs. Post-mitotic, high metabolic demand neurons are highly susceptible to protein homeostasis disruption because of their limited regenerative potential (Table 7.1). Physiological mechanisms of cellular quality control, such as molecular chaperones, the ubiquitin-proteasome system, and autophagy-lysosomal pathways, are responsible of cellular quality control by maintaining appropriate protein folding and clearance of damaged or misfolded proteins. The maladaptation of these systems results in the formation of abnormal protein conformations, which are poisonous to the neuron cells [12]. In conditions of neurodegeneration, proteins undergo conformational change, which facilitates self-association and aggregation. An example is that amyloid-β peptides and hyperphosphorylated tau create extracellular plaques and intracellular neurofibrillary tangles, respectively, in AD. PD and HD are respectively marked by the aggregation of α-synuclein to form Lewy bodies and mutant huntingtin

Table 7.1 Major neurodegenerative diseases and key molecular targets

Disease	*Key pathological features*	*Major molecular targets*
Alzheimer's disease	Amyloid plaques, tau tangles, synaptic loss	BACE1, γ-secretase, Tau, GSK-3β
Parkinson's disease	Dopaminergic neuron loss, Lewy bodies	α-Synuclein, LRRK2, MAO-B
Huntington's disease	Polyglutamine expansion, neuronal death	Mutant HTT, proteostasis regulators
ALS	Motor neuron degeneration, protein inclusions	SOD1, TDP-43, FUS
Frontotemporal dementia	Behavioral and language impairment	Tau, TDP-43

containing extended polyglutamine repeats to intracellular inclusions. Equally, TDP-43 and SOD1 misfolded proteins are also involved in ALS. Such aggregated proteins tend to take on degradation-resistant b-sheet-rich forms, and the aggregates can also be able to seed additional aggregation. Protein aggregates impair the work of neurons in a variety of ways. They disrupt synaptic signaling and axonal transport and modify membrane integrity. Moreover, misfolded proteins may trap such vital cellular components as transcription factors and chaperones, thus worsening cellular stress. Interneuronal transmission of misfolded protein species in a prion-like fashion also increases the progression and propagation of the disease in regions [13,14].

Protein misfolding and aggregation provide feasible targets of intervention, as per a therapeutic perspective. Methods involving stabilization of native protein structure, inhibition of aggregation, increased proteostasis, or clearance of aggregates are under investigation. Computational design strategies are also important in this regard because they allow the identification of aggregation-prone domains, modeling protein folding transitions and screening of small molecules or peptides that can alter aggregation and misfolding pathways. These in silico techniques will be useful in the mechanistic understanding of disease-modifying therapies used to inhibit protein aggregation in neurodegenerative conditions [15,16].

7.2.2 Mitochondrial dysfunction and oxidative stress

Closely interrelated processes include mitochondrial dysfunction and oxidative stress that are leading to the pathogenesis of NDs. The neurons are dependent on oxidative phosphorylation via mitochondria to get their high energy requirements, especially in transmitting and transporting axons and maintaining ionic gradients. In any case, any defect in the functioning of the mitochondria can have extensive effects on the survival and functioning of the neurons [17].

Mitochondria in neurodegenerative disorders have structural and functional defects, such as a deficiency in the activity of the electron transport chain, a lower ATP production rate, mitochondrial dynamics, and calcium homeostasis disruption. These defects cause overproduction of reactive oxygen species (ROS), overloading the cellular antioxidant defenses against them, and causes oxidative stress. Increased concentrations of ROS lead to the destruction of lipids, proteins, and nucleic acids, further weakening neuronal stability and increasing the speed of disease progression (Table 7.1) [18]. Other pathogenic processes related to neurodegeneration are also aggravated by oxidative stress. Oxidative alterations may favor protein misfolding and aggregation, alter impairment of protein degradation via the proteasome, and mitochondrial DNA damage, leading to a self-propagating process of mitochondrial dysfunction and cell damage. Moreover, oxidative stress induces neuroinflammatory pathways and facilitates apoptotic pathways, which are associated with progressive neuronal loss in diseases like Alzheimer disease, PD and ALS. Mitochondrial quality control systems, such as mitophagy and mitochondrial biogenesis, are disrupted in NDs. The role of mitochondrial homeostasis in the health of neurons is evident in defects of major regulatory proteins involved in these processes, including PINK1 and Parkin in PD. Another result of the inability to eliminate degraded mitochondria is dysfunctional organelles, augmentation of oxidative load, and an augmented exposure of neurons [19].

Computational methods offer some good understanding of the mitochondrial dysfunction and oxidative stress-related targets through the possibility of modeling the mitochondrial proteins, redox-sensitive enzymes, and signaling pathways. The in silico studies aid in the discovery of important regulatory nodes, forecasting the occurrence of oxidative stress-related structural modifications in the proteins, and screening of small molecules with antioxidant or mitochondrial protective activity. These calculating plans underpin the rational development of therapeutic interventions that will help to reestablish the mitochondrial functioning and redox equilibrium in NDs [20].

7.2.3 Neuroinflammation and synaptic impairment

Neuroinflammation and synaptic damage are essential factors in the pathogenesis and development of NDs. Although central nervous system inflammation is a protective mechanism in normal circumstances, in chronic situations, it has deleterious consequences on the normal neuronal functions and survival. The sustained neuroinflammatory responses are currently considered a typical pathological feature of major neurodegenerative disorders, such as Alzheimer disease, PD, and ALS. The major effectors of neuroinflammation are microglia and astrocytes. When diseased, these glial cells get chronically activated and release pro-inflammatory cytokines, chemokines, ROS, and nitric oxide. Though the first thing was to remove the pathogenic proteins and cellular debris, as the glial activity is prolonged, the neurons are damaged and the homeostasis of the synapses is destabilized. Inflammatory mediators have the ability to change the neurotransmitters release and damage synaptic plasticity and neuronal communication, thus leading to cognitive impairment and motor impairment [21].

One of the first and most trusted correlates of progression of NDs is synaptic dysfunction. Synapses are very dynamic structures that require accurate control of protein trafficking, energy provision, and calcium signaling. Neuroinflammatory signaling disrupts these pathways by altering the expression of receptors, changing the vesicle cycling in the synaptic area, and interfering with cytoskeleton organization. The inflammatory pathways in the case of Alzheimer pathology increase the vulnerability of the synapses to the toxicity of amyloid-β, and synapses are lost before the actual toxicity of neurons (Table 7.1) [22].

This is because neuroinflammation and synaptic impairment interact in a vicious cycle in neurodegeneration. Damage of synapses may also induce the activation of glial cells, enhancing inflammatory reactions and increasing the pace of the disease. The genetic researchers have reported a number of risk genes related to inflammation, including TREM2 and CD33, which highlights the significance of immune signaling pathways in regulating synaptic integrity and vulnerability to disease [23,24].

Computational design methodologies can be used to systematically investigate neuroinflammatory and synaptic signaling networks using multi-omics data and protein–protein interactions. In silico modeling is useful in identifying major inflammatory mediators and synaptic regulators that can be used as treatment. In addition, computational screening and network-based modeling have facilitated the creation of multi-target approaches to regulate neuroinflammation and maintain synaptic activity at the same time to provide promising opportunities in the area of disease-modulating interventions in NDs [24].

7.3 THERAPEUTIC TARGET LANDSCAPE IN NEURODEGENERATIVE DISEASES

7.3.1 Enzymatic and receptor-based targets

Enzymes and receptors are examples of significant therapeutically actionable molecules in NDs. The targets are specifically appealing to drug discovery because of their clear active or binding sites, predetermined positions in pathological processes, and ability to be manipulated by small molecules. The development of molecular biology and neuropharmacology has revealed many of these enzymes and receptors that form the regulation of several of the most important pathological processes, including protein aggregation, neurotransmission, oxidative stress, and neuroinflammation. There are a number of enzymes that are important in production and clearance of harmful proteins [25]. B-site amyloid precursor protein-cleaving enzyme 1 (BACE1) and g-secretase play a major role in the production of amyloid-b in Alzheimer disease, which is why these two proteins are of prime interest as therapeutic targets. Kinases that are involved in tau hyperphosphorylation and formation of neurofibrillary tangles include glycogen synthase kinase-3b (GSK-3b) and cyclin-dependent kinase 5 (CDK5). Among the enzymatic targets in the pathophysiology of PD, there are monoamine oxidase-B (MAO-B) and leucine-rich repeat kinase 2 (LRRK2), which are both involved in the regulation of dopamine metabolism and neuronal signaling (Figure 7.1) [25].

Figure 7.1 shows the significant pathologic processes that cause NDs and the respective therapeutic target classes. Interplay the processes of interconnected protein misfolding and aggregation, mitochondrial dysfunction and oxidative stress, neuroinflammation, synaptic impairment, and aberrant protein–protein interactions convert into central neuronal degeneration. All these pathological processes result in the different target categories, which include enzymatic and receptor-based targets, protein–protein interaction (PPI) interfaces, and genetic and signaling pathway controls. The schematic emphasizes the multifactorial and complex nature of neurodegeneration

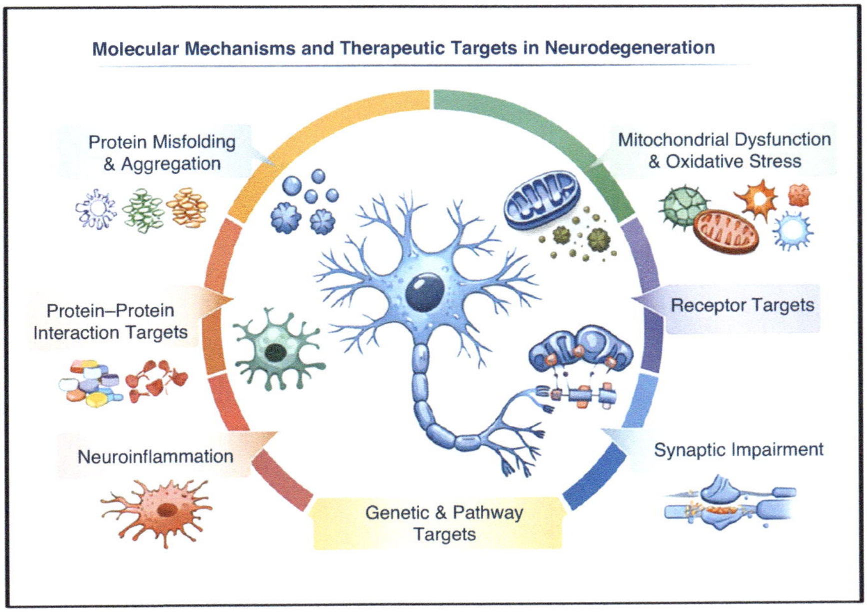

Figure 7.1 Molecular mechanisms underlying neurodegeneration and associated therapeutic targets.

and highlights the need to use multi-target and pathway-based therapeutic approaches. These methods can be mathematically explored through computational design techniques.

Receptors are also significant targets of regulating neuronal communication and survival. Synaptic transmission and plasticity involve directly the neurotransmitter receptors such as *N*-methyl-D-aspartate (NMDA), g-aminobutyric acid (GABA), dopamine, and acetylcholine receptors. Abnormal functioning of these receptors is a cause of excitotoxicity, synaptic failure, and cognitive impairment. Also, neurotrophic factor receptors and immune-related receptors, including toll-like receptors and triggering receptor expressed on myeloid cells 2 (TREM2), are also critical in neuronal maintenance and neuroinflammatory reactions. The enzymatic and receptor-based targets have been greatly identified and optimized as a result of computational design methods. Such techniques as molecular docking and MD simulations are structure-based methods, which allow the active sites and ligand-binding pockets to be characterized in detail, enabling the rational design of inhibitors or agonists. The ligand-based modeling and virtual screening also aid in the discovery of compounds that have high specificity and good pharmacological activities. The strategies are especially effective to maximize BBB permeability and reduce off-target effects. On the whole, enzymatic and receptor-based targets continue to focus on neurodegenerative drug discovery initiatives. The combination of computational design plans with experimental validation schemes enhances the process of target prioritization and lead optimization, thus providing viable perspectives in the formulation of efficient therapeutic interventions that will alter disease progression in NDs [26].

7.3.2 Protein–protein interaction targets

The PPIs are a needful part of cellular homeostasis and central nervous system signaling network regulation. In the NDs, the aberrant PPIs are involved in pathology, including protein aggregation, defective intracellular transport, signaling cascade deregulation, and neuronal death. PPI, in contrast to its enzymatic targets, has large and flat interaction surfaces that are, in most cases, dynamic and traditionally difficult to therapeutically interfere with. Nevertheless, computational design progress has revived the interest to come back to PPIs as a potential intervention for NDs. A number of proteins that are linked with NDs have their pathogenic activity based on aberrant interactions between them and other constituents of the cell [27]. The complex between amyloid precursor protein (APP), secretases, and adaptor proteins in the production and aggregation of amyloid-β is also involved in Alzheimer disease. Tau protein also communicates with the microtubules and other kinases, and when it is disrupted, this disrupts cytoskeletal stability and synaptic defects. The α-synuclein in PD interacts abnormally with membranes, chaperones, and mitochondrial proteins, fostering oligomerization of these toxins and cellular stress. The attack on PPIs provides the prospects of stabilizing the disease mechanisms before the cells have suffered permanent damage. Computational approaches are important in the determination of hotspots of interactions, the mapping of the binding interfaces, and the characterization of transient or weak interactions that are hard to observe experimentally. Computational methods based on structure, such as protein–protein docking and MD simulations, can be used to examine the interface of the interaction and the conformational flexibility in detail. Such approaches assist in determining important residues and regions that are critical in the formation of complexes, which can then be specifically targeted using small molecules, peptides, or peptidomimetics. Computational analyses based on them in a network explicitly augment PPI discovery and target discovery by combining protein interaction data with disease-specific omics data [28]. These methods can be used to identify the hub proteins and bottleneck interactions whose presence in a disease network is disproportionate. Targeting such important nodes of interaction could be more therapeutically effective than acting on individual enzymes or receptors in NDs. In addition, through computational screening of libraries of compounds against PPI interfaces, small-molecule modulators that can disrupt or stabilize discrete interactions and hence change disease-relevant pathways have been discovered. Although there have been these developments, there are still problems in the therapeutic targeting of PPIs such as attaining adequate specificity and brain penetration. There is, therefore, a need to use computational design techniques that integrate optimization of physicochemical properties and prediction of BBB permeability. All in all, PPIs are a potentially beneficial and growing category of ND targets, and computational methods are invaluable tools for their rational recognition and therapeutic exploitation [29].

7.3.3 Genetic and signaling pathway targets

Another important dimension of the therapeutic target space in the pathogenesis of NDs is genetic and signaling pathway targets. Recent developments in genomics and high-throughput sequencing technologies have identified a large number of genetic variants related to disease susceptibility, disease progression, and phenotypic heterogeneity. The genetic findings have identified significant molecular pathways in neuronal survival, protein homeostasis, immune regulation, and synaptic activity, which have served as a basis for therapy-focused pathway interventions. There are a number of neurodegenerative disorders with high genetic elements. Gene mutations that cause familial AD include APP, PSEN1, and PSEN2, while variants of SNCA, LRRK2, and PARK genes are associated with PD. HD is caused by the presence of coded CAG repeats in the HTT gene, and mutations in the SOD1, TARDBP, and FUS genes lead to ALS. These mutations tend to disrupt at overlapping signaling pathways, which involve the processes of autophagy, mitochondrial activity, neuroinflammation, and synaptic plasticity. The use of signaling pathways as targets compared to single-gene products is beneficial in various ways when dealing with NDs. The multifactorial nature of these disorders can be treated using pathway-based interventions, which modulate a number of downstream effectors at the same time [30,31]. It is possible to use computational systems biology methods to reconstruct and analyze disease-specific signaling networks and thereby identify the main regulatory nodes and feedback loops that run pathological processes. These analyses justify preference of targets of centrality in the network and therapeutic interest. Disease progression and treatment response can also be predicted with the assistance of computational modeling of genetic and signaling pathways. Genomic, transcriptomic, and proteomic data are used in integrative approaches, which are able to give information on the dynamic changes in activity of the pathways across the disease. ML algorithms can also be used to discover genotype–phenotype relationships and forecast the functional effects of genetic variants, and these can be used to help select patient-specific or precision medicine targets. Moreover, computational design can aid in the formation of multi-target and combination therapies and simulates the pathway perturbations and evaluates possible synergistic outcomes. This is particularly relevant in NDs, where single-target interventions have failed to produce any significant clinical effects. Computational methods can help identify and characterize robust and translatable targets in the form of therapeutic targets by facilitating rational searches in intricate genetic and signaling landscapes. NDs may be well understood and treated using genetic and signaling pathway targets. Combining computational design with genetic and systems-level approaches will improve the target discovery and personalized therapeutic strategies and give the rational grounds in developing effective disease-modifying interventions [32,33].

7.4 COMPUTATIONAL TARGET IDENTIFICATION AND VALIDATION

Rational drug discovery of NDs is based on the identification and validation of molecular targets. Since these disorders are complex, heterogeneous, and multifactorial, the standard methods of experimental target discovery do not always succeed in the discovery of disease-relevant molecular interactions at the system level. Computational methods provide effective, scalable, and economical approaches to systematically discover, rank, and confirm therapeutic targets by combining and learning across a variety of biological datasets and predictive models [34].

Table 7.2 Computational approaches for target identification and validation [34]

Computational approach	*Purpose*	*Key outcomes*
Sequence-based analysis	Identify conserved and disease-linked regions	Genetically validated targets
Structural annotation	Assess target druggability	Binding site identification
Functional annotation	Assign biological roles	Mechanistic insights
Network biology	Identify hubs and bottlenecks	System-level target prioritization
Systems biology	Integrate multi-omics data	Pathway-level targets

Computational target identification (Table 7.2) allows early-stage determination of relevance, druggability, and safety of the target, which minimizes failures in neurodegenerative drug development in the late phase of development. In silico frameworks can be obtained by integrating sequence-based analyses and structural and functional annotation, as well as network and systems biology, to gain an in-depth view of disease-related molecular processes and allow the rational selection of clinically actionable targets.

7.4.1 Sequence-based target analysis

Target analysis through sequences is one of the main computational approaches to the identification of disease-related genes and proteins. It takes advantage of the information in nucleotide and amino acid sequences to derive an idea of evolutionary preservation, functional significance, and pathological significance [35]. The sequence-based analyses prove useful in NDs in which genetic variations can pose a major determinant of the likelihood and advancement of a disease. Comparative genomics and multiple sequence alignment enable the identification of conserved domains and residues, which are also major in protein functioning. These conserved areas are usually enzyme active sites, ligand-binding motifs, or regulating elements, which make them desirable therapeutic targets. On the other hand, mutations that are related to the disease are often located in these conserved regions and impair the normal functioning of the proteins and cause neurodegenerative pathology. The sequential methods are also commonly applied in the analysis of the pathogenic variants that are detected in genome-wide association studies and next-generation sequencing. The computational methods have the ability to forecast the effect of missense mutations, insertions, deletions, and repeat expansions on protein folding stability and interaction, as well as folding. It is on the basis of this information that genetically validated targets that have a good disease association are prioritized. Besides this, post-translational modification localizations, including phosphorylation, acetylation, ubiquitination, and glycosylation, can also be predicted by sequence analysis and are highly important in neurodegenerative events, such as protein aggregation, signal regulation, and proteolytic pathways. The discovery of intrinsically disordered regions and aggregation-prone sequences is also involved in gaining knowledge about pathogenic processes and finding possible sites of intervention [36].

7.4.2 Structural and functional annotation

Structural and functional annotation will provide important information about the molecular behavior and pharmacological potential of candidate targets. Further developments in protein structure prediction and homology modeling have now greatly increased structural coverage of ND-associated proteins, most of which have no experimentally determined structure because of their dynamism or aggregation-prone character (Figure 7.2) [37]. The task of predicting secondary and tertiary structures, functional domains, and mapping potential ligand-binding pockets or interaction interfaces is referred to as structural annotation. Such analyses are useful in identifying the potential of a protein to be amenable to small-molecule modulation, peptide-based intervention, or biologic targeting. The determination of structural features such as pocket depth, hydrophobicity, and flexibility is a major determinant of drug accessibility and is regularly evaluated with the help of computational tools [38].

Figure 7.2 shows a detailed in silico drug discovery workflow in NDs, which is target-oriented. Multi-omics data (genomics, proteomics, and transcriptomics) are used as the starting point of the pipeline, and they are used to compute target identification. Lead identification is done through structure-based design strategies, for example protein modeling, molecular docking, and ligand-based design strategies, for example, pharmacophore modeling and virtual screening. The methods of AI and ML are helpful to

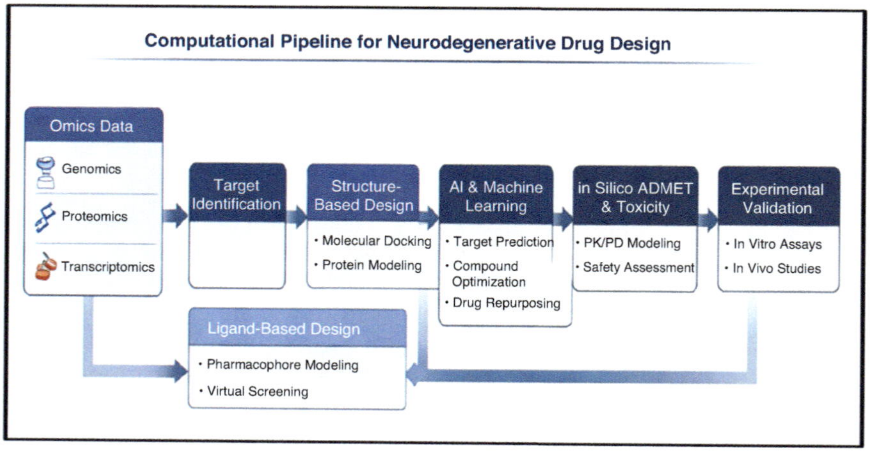

Figure 7.2 Integrated computational pipeline for neurodegenerative drug design.

predict targets, optimize compounds, or repurpose drugs. Lead prioritization is based on in silico ADMET and toxicity tests, such as the pharmacokinetic modeling and safety modeling, before the lead is experimentally validated by in vitro and in vivo experiments.

Structural information is combined with biochemical and biological data in functional annotation to give the molecular functions, enzyme activities, and interaction partners. The functional annotation is also mainly needed in NDs to differentiate between primary drivers of the disease and downstream or compensatory molecules [39]. The potential of structurally complex targets is further increased by computational prediction of allosteric sites and conformational transitions that broadens the therapeutic potential of these complex targets. Also, the rational design of inhibitors, activators, or stabilizers that can be used to regulate the target activity with high specificity is aided by structural and functional annotation. This information underlies further design-based approaches to structures as drug design, such as molecular docking, virtual screening, and MD [40,41].

7.4.3 Network and systems biology approaches

Network and systems biology methods have revolutionized the way targets are identified by allowing the analysis of NDs as complex and interacting systems in biology instead of single molecule events. These methods represent interrelationships between genes, proteins, metabolites, and signaling pathways to reflect the emergent characteristics of disease networks. Construction of PPI networks, gene regulatory network, and signaling pathway maps can be done using computations that identify the most important regulatory nodes, hubs, and bottlenecks that regulate disease-related processes. Targets at the central positions in these networks usually have disproportionate effects on the development of the disease and are promising targets of therapeutic interventions. Multi-omics multi-experimentation systems biology methods also allow the combination of multi-omics datasets with genomics, transcriptomics, proteomics, and metabolomics. Integrative analyses of this nature display disordered pathways and dynamic molecular alterations linked with the onset, progression, and intensity of the disease [42–44]. These networks can be investigated using computational modeling and simulation to predict the effect of altering individual targets to aid sound target validation. Significantly, the network-based approaches are consistent with the new paradigm of ND multi-target and pathway-oriented therapies. Computational systems biology increases the chances of developing effective disease-modifying therapies by identifying viable combinations of synergistic targets that trigger compensatory network responses in a limited way. Computational target discovery and validation is an essential part of contemporary ND drug discovery [45,46]. Through the combination of sequence analysis, structural and functional annotation, and network and systems biology, in silico methodologies offer a holistic and mechanistically informed system of prioritization of therapeutic targets. These approaches can not only make the discovery faster at an early stage but also enhance the translational success of these approaches by matching molecular targets to the complex biology of NDs [47–49].

7.5 STRUCTURE-BASED COMPUTATIONAL DESIGN STRATEGIES

One of the roots of current drug discovery of NDs is structure-based computational design strategies. These methods make use of comprehensive three-dimensional (3D) structural data of biological targets to control the logical development and optimization of therapeutic agents. With the growing access to high-quality structural information and the growth in the power of computing resources, structure-based design, in turn, has become a necessity to aid in the identification of ligands capable of exerting significant control over disease-relevant targets throughout the central nervous system [50].

7.5.1 Protein structure modeling and refinement

Proper structure-based drug designing requires accurate protein structural models. Nonetheless, most ND-related proteins do not have experimentally measured structure because of their inherent disorder, aggregation potential, or conformational plasticity. Computational protein structure modeling will help to get a solution by creating credible 3D structures in terms of sequence data and available structural templates. Homology modeling is common when proteins of interest have structurally related proteins, which makes it possible to build target models by matching the sequence with experimentally resolved templates. In the case of proteins with no close homology, ab initio and hybrid modeling are used to compute plausible conformations. The refinement of these computational models is further worked on with energy minimization and refinement methods based on the use of MD to enhance the accuracy and stability of the structure. The refinement of the structures is especially crucial in the targets of ND, where the small changes in either the conformations of the ligands or the functional results can have a big impact. Refinement protocols maximize the side-chain orientations and steric mismatches as well as enhance the geometry of the backbone, leading to a better representation of target proteins. Structural models that are of high quality are a good starting point in the analysis of binding sites and the design of ligands [51,52].

7.5.2 Binding site prediction and characterization

The ligand-binding sites are important entities that need to be identified and characterized during structure-based drug design. The binding sites can be an active site of an enzyme, a ligand-binding site on a receptor, an allosteric site, or a PPI site. Computational methods of binding site prediction couple protein surface geometry, electrostatic properties, and evolutionary conservation to predict areas of the surface that are likely to accommodate small molecules or peptides. In NDs, characterization of binding sites is usually difficult because of transient and cryptic or conformationally controlled pockets. Computational methods, such as pocket detection methods and dynamic simulations, are useful in revealing hidden or inducible binding sites that otherwise might not be commonly evident in the fixed protein structures [53]. The binding sites are characterized by analyzing the pocket volume, shape, hydrophobicity, and residue composition, which are all used to determine the specificity and druggability of a ligand-binding site. Proper binding site analysis would justify the informed choice of targets and would inform the design of ligands that would have the best interactions. In addition, physicochemical constraint characterization of blood-brain-barrier-relevant can be commonly incorporated at this point to guarantee bioavailability of designed compounds in the central nervous system [54].

7.5.3 Molecular docking approaches

Molecular docking is one of the most important methods of computation that is applied to predict the orientation and affinity of small molecules in the target binding site. Docking algorithms search through the combinations of ligand conformations and positions and determine the strength of interaction between atoms, including hydrogen bonding, hydrophobic contacts, and repulsive forces [55]. The method allows screening vast libraries of compounds in a short period of time with a view to finding a possible lead molecule. Molecular docking has also been widely applied in screening enzyme aggregation inhibitors, neurotransmitter receptor, and protein–protein interface ligands in the context of NDs. Docking studies can offer information on the binding mechanism and can also be used to rank candidate molecules against predicted binding affinity and interaction profiles. Enhanced docking methods include receptor flexibility, solvation, and consensus score so as to have enhanced prediction accuracy. Pharmacokinetic and toxicity prediction: Docking results are most commonly used together with pharmacokinetic and toxicity prediction to focus on compounds that are drug-like [56]. Being a part and parcel of structure-based design, molecular docking allows greatly accelerated early lead discovery and optimization [57].

7.5.4 Molecular dynamics simulations

MD simulations offer a dynamic view of target–ligand interactions in the form of a model of atomic motions with time. In contrast to static docking models, MD simulations have the ability to model conformational flexibility, binding interactions stability, and dynamic structural changes that might happen in a physiological environment. ND targets are of particular interest as MD simulations are useful when the target possesses much conformational variability and is allosterically regulated. Such simulations make possible the measurement of the stability of the ligand-binding complex, detection of critical residues of the interaction, and the examination of conformational transitions of functional states. Free energy calculations using MD techniques increase further the predictions of binding affinity and aid in optimization of leads (Table 7.3) [58].

Moreover, MD simulations are useful in the study of the aggregation processes of proteins and how different small molecules influence aggregation-prone domains. Through the support of the rational design of disease-modifying therapeutics and the provision of mechanistic information at an atomic scale, MD simulations can improve the validity of structure-based drug design and promote the development of disease-modifying therapeutics [59]. The computational design approaches that use structure are relevant toward improved therapeutic discovery of NDs. These methods provide a mechanism-driven design of drugs by precise protein structure modeling, characterization of binding sites, molecular docking, and MD simulations. The combination of these computational methods and experimental validation provides a solid avenue toward the creation of successful and brain-penetrant methods [60].

Table 7.3 Structure- and ligand-based computational design methods

Method	*Category*	*Application in neurodegeneration*
Protein structure modeling	Structure-based	Modeling aggregation-prone proteins
Binding site prediction	Structure-based	Identification of druggable pockets
Molecular docking	Structure-based	Screening inhibitors/ modulators
Molecular dynamics	Structure-based	Studying conformational stability
Pharmacophore modeling	Ligand-based	Identifying key interaction features
QSAR	Ligand-based	Activity prediction and optimization
Virtual screening	Ligand-based	Rapid hit identification

7.6 LIGAND-BASED COMPUTATIONAL DESIGN APPROACHES

Computational design: Ligand-based computational design has a key place in drug discovery in NDs, where high-resolution structural information of the target protein is either limited or unavailable. They are based on the examination of bioactive compounds known for key chemical characteristics that cause the biological activity and on the design of new therapeutic agents. The structural complexity and conformational flexibility of most targets of NDs make ligand-based methods complementary to structure-based methods and highly effective in the process of lead identification and optimization.

7.6.1 Pharmacophore modeling

The basic ligand-based model is pharmacophore modeling, which determines the chemical characteristics necessary to allow a molecule to actually be recognized and be active biologically. A pharmacophore is a hypothetical characterization of a molecular attribute that includes hydrogen bond acceptors and donors, hydrophobic areas, aromatic rings, and charged groups and their spatial arrangement needed to bind a certain biological target (Figure 7.3) [61].

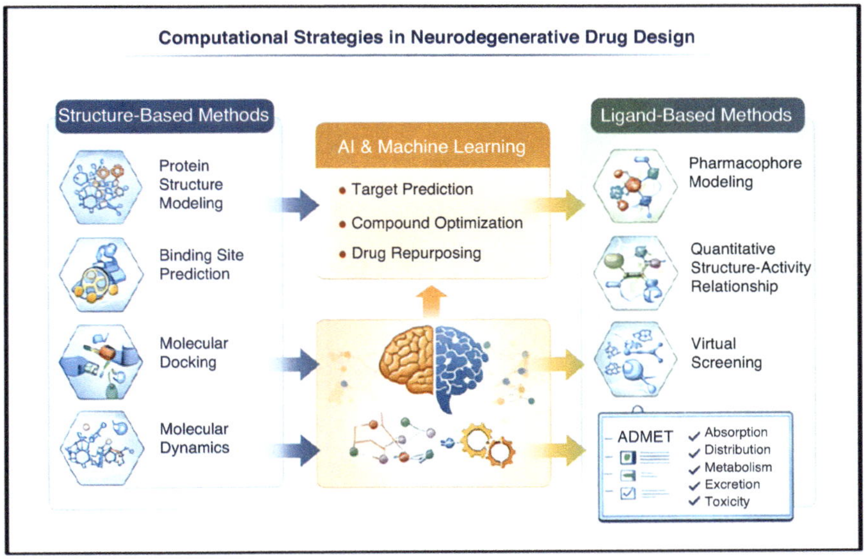

Figure 7.3 Computational strategies for target-oriented drug design in neurodegenerative diseases.

Figure 7.3 is the overview of the key computational strategies that have been used in neurodegenerative drug discovery, including the combination of structure-based, ligand-based, and AI-driven approaches. The modeling of protein structures, the prediction of binding sites, molecular docking, and MD simulations are part of the structure-based approaches, which contribute to the mechanistic understanding of target–ligand interactions. Quantitative structure–activity relationship analysis, pharmacophore modeling, and ligand-based strategies are used to identify and optimize bioactive compounds and include virtual screening. These approaches are filled by ML and AI approaches, which have the capability to predict targets, optimize compounds, and repurpose drugs. Assessment of the pharmacokinetic and safety characteristics in silico points to desirable lead prioritization in neurodegenerative therapeutics.

Pharmacophore models are commonly built in NDs research employing a set of experimentally validated disease-targeted inhibitors or modulators (like enzymes participating in amyloid processing, kinases that participate in tau phosphorylation, or synaptic receptors). These models concentrate on common patterns of interaction between active compounds and could be applied to screen large collections of chemical compounds to discover new molecules that have comparable pharmacological behavior. Pharmacophore modeling is also highly useful to discover compounds that can penetrate the BBB, since physicochemical constraints that apply to activity in the central nervous system may be added to models. Also, a ligand-based pharmacophore may identify new chemical scaffolds that were not previously identified, which facilitates scaffold hopping and chemical diversity in lead discovery. Pharmacophore modeling can be used to complement the discovery of novel neurotherapeutic candidates by concentrating on key aspects of interactions instead of the molecular structure [62].

7.6.2 Quantitative structure–activity relationships (QSAR)

Quantitative structure–activity relationship (QSAR) modeling is a mathematical model of the relationship between the chemical structure of compounds and biological activities. QSAR methods transform molecular properties of the type of hydrophobicity, electronic distribution, steric factors, and topology attributes into numerical values, which are correlated with experimental activity data. The models can be used to predict the effects of untested compounds and do rational optimization of lead molecules. QSAR models have been greatly utilized in the design of aggregation-based enzyme, receptor modulator, and neuroprotective agonist inhibitors in the context of NDs. The methods used to determine structural features identifying potency, selectivity, and central nervous system bioavailability are both classical QSAR and three-dimensional QSAR (comparative molecular field analysis). ML algorithms are being utilized in modern QSAR modeling to process large and complicated datasets. These ML methods enhance the accuracy of prediction and allow the discovery of non-linear correlations between molecular characteristics and biological activity. QSAR models are also important in the lowering of experimental expenses by focusing on compounds with a high predicted efficacy and preferential pharmacokinetic qualities prior to being produced and undergoing biological testing (Table 7.3) [63].

7.6.3 Virtual screening methodologies

Virtual screening: Virtual screening is a high-throughput computational method that is applied to find promising lead compounds among large chemical libraries [64]. In virtual screening using ligands, the compounds are chosen by their structural resemblance of known active molecules or their potential to fulfill pharmacophore models. This method is especially suitable in initial drug discovery in NDs, where it is necessary to investigate chemical space very quickly. Similarity searching, fingerprint-based screening, and pharmacophore-based screening are some of the ligand-based virtual screening techniques. These methods make it possible to conduct effective screening of millions of compounds to find candidates with preferred biological and physicochemical characteristics [65]. Notably, to improve translational applicability, virtual screening processes frequently incorporate central nervous system-specific filters, for example, BBB permeability and prediction of neurotoxicity. Virtual screening techniques are commonly used together with further structure-based techniques, such as molecular docking and MD simulations, to hone selection of hits and confirm binding mechanisms. This combined approach will speed up the process of lead discovery and enhance the chances of discovering compounds that can modify the complicated ND targets. Computational design methods based on ligands are essential in the drug discovery of neurodegradation. These methods can allow the effective identification, prioritization, and optimization of lead compounds, without detailed target structural information, through pharmacophore modeling, QSAR analysis, and virtual screening methodologies [66]. Ligand-based approaches used in conjunction with structure-based and experimental approaches can contribute greatly to the rational design of effective and brain-penetrative therapeutics of ND [67,68].

7.7 ARTIFICIAL INTELLIGENCE AND MACHINE LEARNING IN TARGET-ORIENTED DRUG DESIGN

ML and AI have quickly revolutionized the world of drug discovery by facilitating data-driven, predictive, and automated drug targets and lead optimization discovery. AI and ML provide potent capabilities in enhancing target-oriented drug design in the context of NDs, where the complexity of molecular processes, heterogeneity, and high rates of clinical failure are a major issue. The technologies can be used to identify disease-relevant targets and optimize therapeutic compounds and identify repurposing opportunities with enhanced efficiency and translational potential by integrating a wide range of biological, chemical, and clinical data [69].

7.7.1 Machine learning models for target prediction

The ML models are essential in the prediction and prioritization of therapeutic targets of NDs. Conventional methods of target identification are usually based on manual data interpretation of experimental measurements, which may be restrictive in terms of scope and scalability. ML-based techniques address such shortcomings by acquiring patterns on large-scale sets of data, such as genomic, transcriptomic, proteomic, and PPI data. The algorithms known as supervised learning are typically applied to categorize genes or proteins as possible disease targets depending upon established target properties, including sequence characteristics, expression patterns, network centrality, and evolutionary preservation. Unsupervised learning techniques also allow discovering new clusters of targets based on the identification of hidden structures in multi-omics datasets. These applications are especially useful where NDs are considered where a number of interlinked pathways lead to disease progression [70]. ML is also widely used in the study of genome-wide association studies and rare variations to determine how a genetic change affects biological functions. ML models help to identify genetic validated targets with high causal relationships to neurodegeneration by comparing genetic variants to disease phenotypes and molecular pathways. Notably, such computational predictions aid in ranking targets, which have higher chances of producing disease-modifying effects on therapeutic modulation [71,72].

7.7.2 Deep learning in compound optimization

A branch of ML, deep learning, a type of ML that involves the use of artificial neural networks, including multiple layers of hidden layers, has demonstrated impressive success in compound optimization and molecular design [73]. The complex non-linear relationships between chemical structure and biological activity are in most cases hard to model with conventional computational approaches, but deep learning models can capture them. Deep learning methods are common in neurodegenerative drug discovery to predict small-molecule binding affinity, selectivity, and pharmacokinetic properties. Convolutional neural networks and graph neural networks allow learning directly with molecular structures, which can be used to predict structure–activity relationships with high accuracy [74]. The rationale optimization of lead compounds is assisted with the help of these models which propose structural alterations that improve potency, reduce toxicity, and enhance BBB permeability. Generative modeling, in which new chemical structures are de novo generated based on learned representations of bioactive compounds, is also a use of deep learning. These generative methods are capable of expanding the chemical search space and identifying novel scaffolds that can be utilized to achieve new targets in NDs, such as PPI interfaces, aggregation-prone proteins, and others. Deep learning can save a large amount of time and money used in lead optimization because it shortens the iterative design cycle [75].

7.7.3 AI-assisted drug repurposing

A promising area of application of AIs in drug repurposing has been in NDs where the creation of new drugs is slow and has a high rate of failure. Drug repurposing refers to the process of finding new therapeutic uses of existing drugs with regard to their prior safety records as well as clinical information. This process is improved by using AI and ML approaches to systematically analyze large-scale biomedical data and reveal concealed interactions between drugs, targets, and diseases. ML models combine data based on gene expression, disease signature, chemical structure, and clinical outcomes to identify drugs that could be used to regulate ND pathways [76]. AI schemes that rely on networks also determine drugs that have the potential to destabilize disease-related molecular networks and regain system-wide stability. These techniques have been useful in screening repurposing candidates of Alzheimer's, Parkinson's, and ALS diseases. The multi-target and pathway-based therapeutic approach is also beneficial to the targeting of challenging neurodegenerative pathways through AI-assisted repurposing. Computational repurposing strategies can shorten the duration of translational research and decrease the risk of late-stage clinical development because the drugs they focus on have good central nervous system exposure and low toxicity [77].

7.7.4 Integration of AI with computational drug design pipelines

The combination of AI and ML and the existing computational drug design pipelines has added to their influence on the research of NDs. Target prediction models based on AI assist in structure-based and ligand-based design: they enable the prediction of high-value targets and pertinent binding sites [78]. The scoring functions that are based on deep learning can enhance the accuracy of molecular docking, and the supercomputing of binding affinity can be obtained using ML-enhanced MD. Besides, AI makes it easier to combine multi-omics and longitudinal clinical data and dynamically model the course of the disease and treatment response. These integrative strategies can be used to uphold precision medicine strategies, that is, by establishing patient-specific targets and optimizing therapeutic interventions through a patient-specific molecular profile. In spite of these developments, issues in the use of AI and ML are problems in data quality, model interpretability, and generalizability. The datasets of NDs tend to be heterogeneous and small, which may have an impact on the robustness of models [79]. To resolve these issues, great care should be taken in model validation, algorithm design, and close interactions with experimental and clinical research. ML and AI are now essential elements of target-based drug design for NDs. These technologies can be used to optimize neurotherapeutic discovery and improve its efficiency, accuracy, and translational relevance by predicting targets more efficiently and optimally with the application of deep learning; help in drug repurposing with AI assistance; and, more importantly, can be used to optimize drug repurposing. Combined with structure-based, ligand-based, and experimental methods, AI-driven strategies provide a useful framework to overcome the difficulties of neurodegeneration and speed up the creation of effective disease-modifying therapy [80].

7.8 ARTIFICIAL INTELLIGENCE AND MACHINE LEARNING IN TARGET-ORIENTED DRUG DESIGN

AI and ML have quickly revolutionized the drug discovery field by facilitating data-driven predictive and automated methods of target identification and optimization of leads. AI and ML can be useful in enhancing target-oriented drug design within the context of NDs, where the complex molecular processes, heterogeneity, and high clinical failure rates are challenging. Using a combination of different biological, chemical, and clinical data, the technologies have the ability to identify disease-relevant targets, optimize therapeutic compounds, and find repurposing opportunities with greater efficiency and translational potential [81,82].

7.8.1 Machine learning models for target prediction

The ML models are essential in the prediction and prioritization of therapeutic targets of NDs. Conventional methods of target identification are usually based on manual data interpretation of experimental measurements, which may be restrictive in terms of scope and scalability [83]. ML-based techniques address such shortcomings by acquiring patterns on large-scale sets of data, such as genomic, transcriptomic, proteomic, and PPI data. The algorithms known as supervised learning are typically applied to categorize genes or proteins as possible disease targets depending upon established target properties, including sequence characteristics, expression patterns, network centrality, and evolutionary preservation. Unsupervised learning techniques also allow discovering new clusters of targets based on the identification of hidden structures in multi-omics datasets. These applications are especially useful where NDs are considered, where a number of interlinked pathways lead to disease progression. ML is also widely used in the study of genome-wide association studies and rare variations to determine how a genetic change affects biological functions [84].

7.8.2 Deep learning in compound optimization

A branch of ML, deep learning (which uses artificial neural networks with multiple hidden layers) has demonstrated impressive performance in optimization and design of compounds. The models of deep learning have the ability to represent highly non-linear relationships between chemical structure and biological activity that are commonly challenging to model using conventional computational approaches. Deep learning is a popular method of neurodegenerative drug discovery, which involves predicting the binding affinity, selectivity, and pharmacokinetics of small molecules [85,86]. Convolutional neural networks and graph neural networks provide the opportunity to learn directly based on molecular structures and predict structure–activity relations. The following models are beneficial to rational optimization of lead compounds, which propose structural changes such that potency, toxicity, and BBB permeability can be improved. Generative modeling is also addressed using deep learning, in which new chemical structures are designed in de novo using learned representations of bioactive compounds [87]. These types of generative methods increase chemical search space and enable the identification of novel scaffolds that can be used to attack difficult ND targets, such as PPI interfaces and aggregation-prone proteins. Deep learning saves time and cost lead optimization since it speeds up the iterative development path [88].

7.8.3 AI-assisted drug repurposing

The latest finding in using AI to aid in drug repurposing has been in NDs because pharmaceutical development of new drugs is slow and linked with a high failure rate. Drug repurposing entails the discovery of new therapeutic uses of existing drugs by exploiting the existing safety profiles and clinical data. AI and ML technologies can improve the process by systematically processing high-volume biomedical data to reveal previously undetected relationships between drugs, targets, and diseases. Data on gene expression profiles, disease signatures, chemical structures, and clinical outcomes are integrated using ML models to predict drugs that can be used to regulate ND pathways [89]. The network-based AI methods also proceed to discover the drugs capable of destabilizing disease-related molecular networks and regaining balance on the system level. These techniques have been useful in the determination of repurposing candidates to AD, PD, and ALS. The repurposing of AIs is particularly beneficial in targeting complex neurodegenerative pathways since it helps in fostering multi-target and pathway-based therapeutic approaches. Computational repurposing methods enable faster translational studies by focusing on pharmaceuticals with good central nervous system exposure and low toxicity, and low the risk of late-stage clinical discontinuation [90].

7.8.4 Integration of AI with computational drug design pipelines

The incorporation of AI and ML in previously known pipelines of computational drug design has added additional value to their contribution to research on NDs. Predictive models of AI-enhanced target prediction are used to direct the structure-based and ligand-based design approaches by ranking high-value targets and binding sites [91]. The scoring functions of deep learning are more accurate in molecular docking, whereas the analysis of MD with the help of ML is more efficient in estimating the binding affinity. In addition, AI allows uniting multi-omics and longitudinal clinical data, allowing the dynamic modeling of disease progression and response to the treatment [92]. These integrative methods help accuracy medicine programs by discovering patient-specific destinations and maximizing treatments interventions, using individual molecular profiles (Table 7.4). In spite of these developments, some issues still persist with the use of AI and ML, as regards the quality of data, the interpretability of the model and its generalizability. The datasets of NDs are frequently not homogeneous and are of small size, which may influence the model robustness. To overcome these issues, it is necessary to pay close attention to model validation, clear algorithm design, and a close relation with experimental and clinical research [93]. ML and AI are now inseparable factors of target-based designing of drugs in neuro degenerative disorders. These technologies have optimized the power, precision, and translatability of neurotherapeutic discovery by improving efficiency, precision, and translatability through deep learning-enhanced compound optimization, and AI-assisted drug repurposing.

Table 7.4 Role of AI and in silico tools across drug discovery stages

Drug discovery stage	*In silico tool*	*Impact*
Target prediction	Machine learning	Identification of disease-relevant targets
Lead identification	Virtual screening	Rapid hit discovery
Lead optimization	Deep learning, QSAR	Improved potency and selectivity
Drug repurposing	AI-based network analysis	Faster translational potential
Safety assessment	ADMET prediction models	Reduced late-stage failure

Combined with structure-based, ligand-based, and experimental strategies, AI-driven strategies can be utilized as a solid framework to tackle the challenges of neurodegeneration and hasten the process of creating effective disease-modifying therapies [94].

7.9 IN SILICO PHARMACOKINETICS AND SAFETY ASSESSMENT

Not only the efficacy of therapeutic agents in removing molecular targets but also the capacity of such agents to reach the central nervous system in a safe and therapeutically relevant concentration is of importance to the success of therapeutic agents against NDs. Most of the neurotherapeutic candidates that are promising fail in preclinical or clinical development because of improper pharmacokinetic characteristics or unbearable toxicity [95]. In silico pharmacokinetics and safety testing offer a vital instrument on early-stage examination of drug-like qualities to allow the prioritization of compounds that have good absorption, circulation, metabolism, excretion, and poisonousness (ADMET). The computational pharmacokinetic and safety prediction methods can decrease experimental dropouts, decrease resources used, and inform rational optimization of compounds. This is especially important in the drug discovery of NDs, in which the exposure of the central nervous system and long-term safety are most important [96].

7.9.1 Blood-brain barrier permeability prediction

The BBB is a very selective physiological barrier that limits the access of most substances to the brain and this is a significant challenge to ND therapeutics. The effective drug candidates should have physicochemical properties that would allow adequate penetration of the drug into the BBB without compromising specificity and safety of the target. In silico BBB permeability prediction models are very important in determining compounds that have access to the central nervous system during the initial stages of drug discovery [97–99]. Computational methods of predicting BBB include molecular descriptors like lipophilicity, molecular weight, hydrogen bond ability, polar surface area, and ionization state. Statistical models trained on experimentally verified BBB-permeable and non-permeable compounds can be rapidly used to classify and estimate the ability to penetrate the brain, based on ML. These models are useful in eliminating compounds which have poor CNS exposure before expensive syntheses and in vivo testing. Also, interactions with efflux transporters like P-glycoprotein that actively eliminate compounds in the brain can be predicted with computational techniques. The determination and reduction of efflux liability by in silico analysis increase the chances of achieving long-term therapeutic levels in the central nervous system. BBB prediction, therefore, is an important filter during the design of ND therapeutics [100,101].

7.9.2 ADMET profiling

ADMET profiling refers to the forecasting of the absorption, distribution, metabolism, excretion, and toxicity of an individual substance which together defines the pharmacokinetic behavior of a compound and its clinical viability. In silico ADMET models are models that combine physicochemical descriptors, structural descriptors, and experimental data to assess drug-likeness and detect potential liabilities at an early phase of drug development (Table 7.4) [102]. In the context of drug discovery on NDs, the predictive pharmacokinetic parameters are oral bioavailability, plasma protein binding, and tissue distribution, including the brain exposure. Prediction models on metabolism measures the interaction by metabolic enzymes, specifically cytochrome P450 isoforms to predict metabolic stability and drug–drug interactions. Predictions of excretion also inform dose and clearance route choices of the compound.

Figure 7.4 demonstrates some of the important strategies employed in the computations that are applied to test the pharmacokinetic behavior and safety profiles of candidate neurotherapeutics. The key element in workflow is the ADMET profiling, which includes the aspects of absorption, distribution, metabolism, excretion, and toxicity prediction. The analysis of BBB permeability is highlighted as one of the key factors that determine exposure of drugs to the central nervous system. Metabolism prediction and toxicity prediction cytotransformation biotransformation mediated by cytochrome P450 Computational models are used to predict the potential safety liabilities at the earlier phases of drug discovery. Such in silico

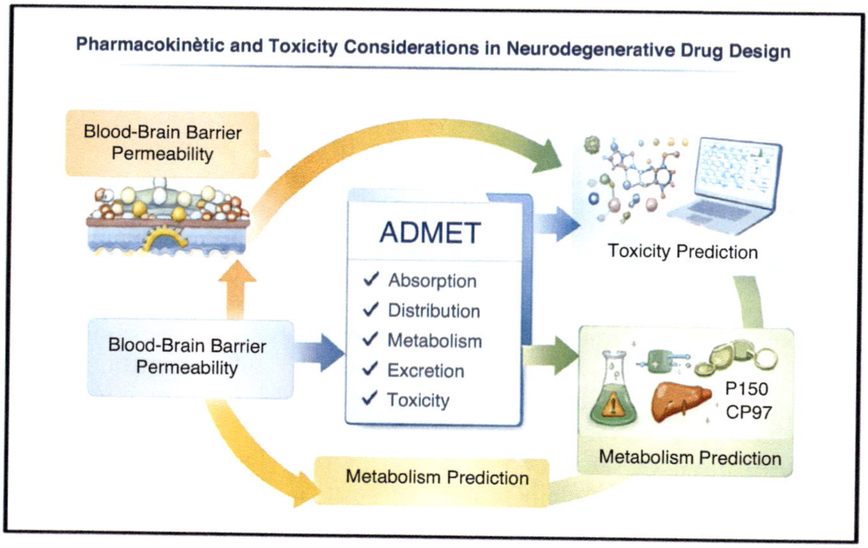

Figure 7.4 In silico pharmacokinetic and toxicity assessment in neurodegenerative drug design.

assessments are used to rationalize lead selection and minimize dying in phase II neurodegenerative therapeutic development [103,104].

ADMET profiling using computational techniques allows a quick screening of a series of lead candidates and rational selection of pharmacokinetics. With early detection of the suboptimal features, researchers are able to alter chemical structures to enhance stability, bioavailability and CNS exposure. Assimilating the ADMET forecasts into calculation lines of design allows to boost the efficiency and translatability of neurotherapeutic development (Figure 7.4) [105].

7.9.3 Toxicity and off-target prediction

Drug development is a significant area of safety assessment, especially NDs, which can be treated over time or on a chronic basis. Models of in silico toxicity prediction assess the risk of possible adverse effects (such as hepatotoxicity, cardiotoxicity, genotoxicity, and neurotoxicity) according to chemical structure and previously identified toxicological patterns. Computational methods also indicate off-target events, which could cause undesirable side effects or a loss of therapeutic efficacy. In silico tools can be used to reveal proteins and pathways that can be unintentionally perturbed by candidate compounds by comparing their structural similarity, compatibility of their binding sites, and interactions at a network level. This type of predictions is necessary to reduce the negative effects and enhance selectivity [106]. The safety is also assessed with the help of ML and network-based toxicity models that are built on a variety of datasets, such as clinical adverse event reports and molecular interaction networks. These methods help in the discovery of compounds of high risk and helps in the further structural optimization to minimize toxicity and off-target behavior. Computational safety assessment in the context of NDs has a critical role in the development of safe and effective therapeutics because the patient is particularly vulnerable in this type of disease [107]. The in silico analysis of pharmacokinetics and safety is an unavoidable part of the drug discovery process in NDs. Computational methods allow one to identify viable therapeutic candidates with an ideal CNS exposure and safety profile early through BBB permeability prediction, comprehensive ADMET profiling, and toxicity and off-target analysis. Combination of these predictive approaches with structure-based, ligand-based, and AI-based design tactics greatly enhances the chances of success in the clinic in the development of neurodegenerative drugs [108].

7.10 CASE STUDIES IN COMPUTATIONAL TARGET DESIGN

Case studies are a vital source of knowledge regarding the application of computational design strategy in the context of ND targets. Computational methods have played a significant role in target identification, validation and lead optimization by combining bioinformatics, structure-based modeling, ligand-based models as well as AI. In this section, the representative case studies of the AD, Parkinson's disease, and emerging computationally identified targets are highlighted, showing the potential of the translational capabilities of the in silico methods of neurodegenerative drug discovery [109].

7.10.1 Alzheimer's disease

The most common ND and one that has been widely investigated using a computational method is that of AD. It is a complex disease with unmet therapeutic requirements. Although over 900 protein targets have been pursued as baits to inhibit AD, the proteins targeted by computer-aided design have been those implicated in amyloid-b (Ab) production and aggregation, tau hyperphosphorylation, synaptic dysfunction, and neuroinflammation [110]. Among the finest examples of a computational target design in AD, there is the example of b-site amyloid precursor protein-cleaving enzyme 1 (BACE1), which is a major enzyme that produces Ab. The BACE1 active site has been characterized and potent inhibitors identified by a number of structure-based computational techniques such as homology modeling, molecular docking, and MD simulations [111]. Large screening of compound libraries have been enabled by virtual screening as lead candidates with high binding affinity and enhanced selectivity are easily identified. Further optimization of chemical scaffolds was directed by in silico optimization to increase BBB permeability and metabolic stability. Computational investigation has also concentrated a lot on tau protein. Considering its intrinsically disordered character, computational modeling and MD simulations were useful in defining areas of aggregation-prone regions and conformational transition related to tau pathology [112]. Virtual screening and ligand-based pharmacophore modeling have been used to identify small molecules capable of inhibiting tau aggregation or controlling the tau phosphorylation pathway. These computational studies provided a mechanistic understanding of tau pathology and guided an experimental validation process. Moreover, network-based methods of computation have been used to discover new AD targets other than traditional amyloid and tau pathways. Transcriptomic and proteomic analyses of systems biology have shown dysregulated signaling networks of synaptic and immune regulation and lipid metabolism. These studies suggest the importance of computational target design in the discovery of multi-target and pathway-directed therapeutic interventions in AD [113].

7.10.2 Parkinson's disease

PD is an ailment that leads to the slow death of dopaminergic neurons, coupled with a-synuclein aggregation. Computational methods have been instrumental in the understanding of molecular pathways that contribute to PD and finding therapeutic targets of protein aggregation, mitochondrial impairment, and neuroinflammation. Computational modeling of a-Synuclein has been widely done since α-Synuclein has an intrinsically disordered structure and is prone to aggregation [114]. The simulation of MD and conformational ensemble has made it possible to characterize a-synuclein folding pathways and the oligomerization mechanisms. These lessons have been used in designing small molecules and peptides that are able to stabilize non-toxic conformations or prevent aggregation. Virtual screening using ligands has also enabled the discovery of agents that mediate a-synuclein-membrane and α-synuclein–chaperone protein interaction [115]. Kinase targets related to familial and sporadic PD have also been being treated with computational target design. The leucine-rich repeat kinase 2 (LRRK2) has become a well-known target therapeutic and structure-based modeling has been utilized to develop selective kinase-inhibitors. Optimization of the inhibitor binding and minimization of the off-target effects have been optimized by docking and MD simulations. Such computational approaches have helped in fastening the optimization of leads and guiding preclinical development [116].

7.10.3 Emerging computationally identified targets

In addition to the established ND targets, computer techniques have also enabled the identification of the emerging targets that have high therapeutic potential. The developments in the AI field, ML, and multi-omics integration have facilitated the detection of novel molecular candidates that were not previously considered through the traditional experimental technology [117]. Proteins that regulate ND progression have been identified through network-based target identification as proteins that regulate autophagy, endosomal trafficking, and lipid metabolism. Computationally simulated PPI networks have revealed hub proteins and bottlenecks that have a strong control over disease-related pathways. These nodes are targeted, which provide prospects of multi-target modulation and disease modification [118]. AI drug repurposing has also been used to identify new targets by discovering new relationships between current drugs and ND pathways. The incorporation of clinical data with gene expression signatures into ML models has made drugs that can regulate disease-relevant targets of interest in order to enable rapid translational discovery [119]. Moreover, the computational prediction of the genetic and epigenetic regulators has identified new targets of transcriptional regulation, RNA processing, and chromatin remodeling. These are the emerging targets that broaden the therapeutic space and assist in highlighting the changing role of computational design in revealing innovative intervention strategies of NDs. These case studies show how important the computational target design is to research in NDs. Since classical targets in Alzheimer and PDs to new computationally determined targets, in silico methods have enhanced target discovery, contributed to better mechanistic insight and rational therapeutic development [120]. The combination of computational strategies and experimental validation remains a source of innovation and has great potential as to the development of effective disease-modifying therapies in the neurodegenerative disorders.

7.11 CHALLENGES, LIMITATIONS, AND FUTURE DIRECTIONS

Irrespective of tremendous progress in the field of computational design and its increasing involvement in studying NDs, there are a number of challenges and limitations. To get the full potential of in silico methods, it is necessary to address these problems to provide the most effective, disease-modifying neurotherapeutics. In this section, the important computation and biological limitations and the new multi-target and network-based approaches and the future will be discussed toward precision medicine.

7.11.1 Computational and biological constraints

The complexity of disease biology presents one of the most significant issues in the design of computational targets in the case of the NDs. Neurodegeneration is dynamic and multifactorial, as it entails genetic, epigenetic, environmental, and age aspects. It is a complicated phenomenon that is hard to capture in computational models due to the use of simplified assumptions or atypical datasets in many models. Computationally, it is seen that the quality and availability of biological data determine a lot in model reliability. ND data tend to be heterogeneous, small, or skewed to particular disease phases or groups of people. The partial structural data, especially of intrinsically disordered and aggregation-prone proteins, also limit the accuracy of structure-based modeling [121]. Additionally, not all computational predictions have been experimentally validated, so their translational relevance is restricted. Biological limits are also a great challenge. Compensatory mechanisms associated with redundancy in cellular networks can make single-target interventions that are predicted computationally less effective. Moreover, variations in in vitro systems, animal systems, and the biology of human disease complicate the validation and explanation of the computational results. To address these limitations, it is necessary to have better data fusion, more powerful validation pipelines, and better correspondence between computational forecasts and experimental design [122].

7.11.2 Multi-target and network-based therapeutic design

Conventional drug discovery models have been mainly one-target based but this approach has proven to be not very successful in NDs. The network of interrelated disease-related pathways implies that successful therapeutic intervention may require simultaneous control of many targets or network elements. Multi-target and network-based therapeutic design has thus become a promising perspective for the future. Disease-relevant molecular networks can be identified using computational approaches in network biology to identify important nodes, hubs, and feedback loops in molecular networks [123]. By manipulating these components of the network, therapeutics will have a wider system-level influence and may circumvent compensatory responses. Computational modeling also aids in the logical design of combination interventions and polypharmacological agents that can regulate various disease pathways simultaneously. Despite this promise, multi-target strategies cause further complexity in drug design, safety assessment and clinical development [124]. Computational tools are important in efficacy and toxicity balancing by predicting the synergies and reducing the adverse interactions. Further development of network models and more knowledge of disease-specific molecular interactions will be needed to realize multi-target strategies as a translatable therapeutic approach.

7.11.3 Precision medicine perspectives

Precision medicine is a new revolution in the future treatment of NDs that seeks to provide a therapeutic approach specific to patient characteristics. Developments in the field of genomics, transcriptomics, and clinical data gathering have indicated a great variation in inter-individual differences in disease mechanisms, course of a disease, and response to treatment. The computational approaches are in a unique position that would enable the combination of these varied data sources and aid in targeted identification and designing of drugs on a personalized basis. Models in ML and systems biology have the ability to stratify patients using molecular signatures, predict disease progression and discover patient-specific treatment targets [125]. This would allow the designing of specific interventions that consider the genetic background, subtype of disease, and progression stage. Computational precision medicine strategies can be important in improving clinical outcomes in NDs, where early intervention is essential [126]. Notwithstanding, the introduction of precision medicine is associated with issues of data availability, data standardization, and ethical aspects. Big, longitudinal data are needed to construct strong predictive models and close-guarded validation is needed to ensure clinical reliability. With a solution to these challenges, computational precision medicine is likely to become part of the core of the ND therapeutics. Although computational design has already revolutionized neurodegenerative drug discovery, it will reach its full potential by enhancing data integration, network-level reasoning, and person-centered treatment. Further methodological development and close interaction among computational scientists, biologists, and clinicians will be required to transform the knowledge gained through computational approaches into therapeutic interventions with potential in treating neurodegenerative disorders [127,128].

7.12 CONCLUSION

The fast development of computational design practices has dramatically changed the environment of NDs research and therapeutic design. Throughout this chapter, it becomes clear that in silico methods have evolved to become supportive tools for target identification, validation, and drug discovery. Computational frameworks have been developed with the aid of sequence-based analysis, structural and functional annotation, ligand- and structure-based design strategies, systems biology, and AI to provide a more comprehensive and mechanistic view of the biology of NDs. Among the most remarkable developments mentioned in this chapter, the possibility of computational methods to deal with the complex nature of neurodegenerative disorders can be singled out. Other diseases like Alzheimer's and Parkinson's are not caused by isolated abnormalities of single molecules but rather by complex pathological mechanisms related to protein aggregation, mitochondrial dysfunction, neuroinflammation, synaptic failure, and genetic vulnerability [129].

In silico methodologies have been used to make drug discovery pipelines more efficient and less prone to attrition. Rational exploration of interactions between targets and ligands can be done by structure-based methods, including protein modeling, molecular docking and MD simulations before experimental experiments. Ligand-based methods, such as pharmacophore modeling, QSAR and virtual screening, can be effective in identifying and optimizing lead compounds within a short period with little structural information available. Combining these strategies will decrease cost, shorten development timelines and enhance the chances of finding candidates with good central nervous system exposure and safety profiles. AI and ML are some of the significant advances in computational neurotherapeutics. Computational design can now be extended to predicting targets due to AI-based target prediction, deep learning-based compound optimization, and data-guided drug repurposing. These methods are especially useful in NDs, which have high failure rates and modest success in developing new drugs that modify the disease. This has hindered traditional drug development. Through the use of massive biological and chemical data, AI-driven solutions can reveal hidden trends, focus on high-value targets, and present novel therapeutic interventions that may not be visible to traditional methods [5]. In the near future, in silico design in neurodegenerative therapeutics is projected to continue increasing in importance. One of the directions is the shift toward multi-target and network-based therapeutic design as single-target strategies have been abandoned. Computational systems biology has shown that the modulation of critical nodes or pathways in networks can provide more stable and long-term therapeutic outcomes compared to the modulation of single molecules. These strategies are very much in line with the biology of NDs that are complex and adaptive. Another significant future trend is the development of precision medicine. Genomic, transcriptomic, proteomic, and clinical data computational integration will allow the stratification of patients and the discovery of personalized therapeutic targets. The progression of diseases, the response to the treatment, and the best intervention strategies can be predicted by ML models and used to develop personalized neurotherapeutics. With the increasing accessibility of longitudinal patient datasets and actual clinical data in real-world settings, in silico methods of precision medicine will have a major impact on the design of clinical trials and positive patient outcomes. Regardless of these advances, further development will be based on overcoming the current limits, such as data heterogeneity, model interpretability, and translational validation. Intimate collaborations between computational scientists, experimental biologists and clinicians will be necessary to ensure that computational predictions have biological significance and clinical relevance [130]. The next generation of neurodegenerative drug discovery will be characterized by the combination of in silico methods, experimental validation, and a clinical feedback loop. Computational design could now be regarded as an essential part of ND research. Its further growth and assimilation with experimental and clinical sciences have tremendous potential in hastening the process of discovering safe, effective, and disease-modifying therapeutics. With the growing capabilities of computational tools, data resources, and interdisciplinary collaboration, in silico design is set to contribute to the solution of the global burden of NDs in a transformative way.

REFERENCES

1. Lamptey RNL, Chaulagain B, Trivedi R, Gothwal A, Layek B, Singh J. A review of the common neurodegenerative disorders: current therapeutic approaches and the potential role of nanotherapeutics. *Int J Mol Sci.* 2022 Feb 6;23(3):1851. doi: 10.3390/ijms23031851. PMID: 35163773; PMCID: PMC8837071.
2. Sheikh S, Safia, Haque E, Mir SS. Neurodegenerative diseases: multifactorial conformational diseases and their therapeutic interventions. *J Neurodegener Dis.* 2013;2013:563481. doi: 10.1155/2013/563481. Epub 2012 Dec 30. PMID: 26316993; PMCID: PMC4437348.
3. Guo F, Liu X, Cai H, Le W. Autophagy in neurodegenerative diseases: pathogenesis and therapy. *Brain Pathol.* 2018 Jan;28(1):3–13. doi: 10.1111/bpa.12545. Epub 2017 Aug 6. PMID: 28703923; PMCID: PMC5739982.

4. Haramshahi SMA, Hamblin MR, Ravesh RK, Sadr H, Ahmadirad N, Mehrabi F, et al. Biomaterials for CNS disorders: a review of development from traditional methods to AI-assisted optimization. *J Mater Sci Mater Med.* 2025;36:85. doi: 10.1007/s10856-025-06947-7
5. Agamah FE, Mazandu GK, Hassan R, Bope CD, Thomford NE, Ghansah A, et al. Computational/in silico methods in drug target and lead prediction. *Brief Bioinform.* 2020 Sep 25;21(5):1663–1675. doi: 10.1093/bib/bbz103. PMID: 31711157; PMCID: PMC7673338.
6. Balch WE, Morimoto RI, Dillin A, Kelly JW. Adapting proteostasis for disease intervention. *Science.* 2008;319(5865):916–919. doi: 10.1126/science.1141448.
7. Pegoraro G, Voss TC, Martin SE, Tuzmen P, Guha R, Misteli T. Identification of mammalian protein quality control factors by high-throughput cellular imaging. *PLoS One.* 2012;7(2).
8. Morris JH, Meng EC, Ferrin TE. Computational tools for the interactive exploration of proteomic and structural data. *Mol Cell Proteomics.* 2010 Aug;9(8):1703–1715. doi: 10.1074/mcp.R000007-MCP201. Epub 2010 Jun 4. PMID: 20525940; PMCID: PMC2938059.
9. Kühner S, van Noort V, Betts MJ, Leo-Macias A, Batisse C, Rode M, et al. Proteome organization in a genome-reduced bacterium. *Science*, 2009;326:1235–1240.
10. Kim PM, Lu LJ, Xia Y, Gerstein MB. Relating three-dimensional structures to protein networks provides evolutionary insights. *Science.* 2006;314:1938–1941.
11. Ferreira LG, Dos Santos RN, Oliva G, Andricopulo AD. Molecular docking and structure-based drug design strategies. *Molecules.* 2015 Jul 22;20(7):13384–13421. doi: 10.3390/molecules200713384. PMID: 26205061; PMCID: PMC6332083.
12. Koszła O, Sołek P. Misfolding and aggregation in neurodegenerative diseases: protein quality control machinery as potential therapeutic clearance pathways. *Cell Commun Signal.* 2024 Aug 30;22(1):421. doi: 10.1186/s12964-024-01791-8. PMID: 39215343; PMCID: PMC11365204.
13. Yang J, Zhi W, Wang L. Role of tau protein in neurodegenerative diseases and development of its targeted drugs: a literature review. *Molecules.* 2024 Jun 13;29(12):2812. doi: 10.3390/molecules29122812. PMID: 38930877; PMCID: PMC11206543.
14. Goedert M, Spillantini MG, Jakes R, Rutherford D, Crowther RA. Multiple isoforms of human microtubule-associated protein tau: sequences and localization in neurofibrillary tangles of Alzheimer's disease. *Neuron.* 1989;3:519–526. doi: 10.1016/0896–6273(89)90210-9.
15. Mandelkow EM, Mandelkow E. Biochemistry and cell biology of tau protein in neurofibrillary degeneration. *Cold Spring Harb Perspect Med.* 2012;2:a006247. doi: 10.1101/cshperspect.a006247.
16. Waheed Z, Choudhary J, Jatala FH, Fatimah NA, Zerr I, Zafar S. The role of tau proteoforms in health and disease. *Mol Neurobiol.* 2023;60:5155–5166. doi: 10.1007/s12035-023-03387-8.
17. Klemmensen MM, Borrowman SH, Pearce C, Pyles B, Chandra B. Mitochondrial dysfunction in neurodegenerative disorders. *Neurotherapeutics.* 2024 Jan;21(1):e00292. doi: 10.1016/j.neurot.2023.10.002. Epub 2023 Dec 19. PMID: 38241161; PMCID: PMC10903104.

18. Ballard JW, Whitlock MC. The incomplete natural history of mitochondria. *Mol Ecol.* 2004;13:729–744. doi: 10.1046/j.1365-294x.2003.02063.x
19. Brezic N, Gligorevic S, Sic A, Knezevic NN. Protein misfolding and aggregation as a mechanistic link between chronic pain and neurodegenerative diseases. *Curr Issues Mol Biol.* 2025 Apr 8;47(4):259. doi: 10.3390/cimb47040259. PMID: 40699658; PMCID: PMC12026403.
20. Kumar R, Jafri MS. Computational modeling of mitochondria to understand the dynamics of oxidative stress. *Methods Mol Biol.* 2022;2497:363–422. doi: 10.1007/978-1-0716–2309-1_27. PMID: 35771458; PMCID: PMC9811848.
21. Adamu A, Li S, Gao F, Xue G. The role of neuroinflammation in neurodegenerative diseases: current understanding and future therapeutic targets. *Front Aging Neurosci.* 2024 Apr 12;16:1347987. doi: 10.3389/fnagi.2024.1347987. PMID: 38681666; PMCID: PMC11045904.
22. Aboudounya M, Heads R. COVID-19 and toll-like receptor 4 (TLR4): SARS-CoV-2 may bind and activate TLR4 to increase ACE2 expression, facilitating entry and causing hyperinflammation. *Mediat Inflamm.* 2021;2021:8874339. doi: 10.1155/2021/8874339
23. Abdelhamid M, Zhou C, Ohno K, Kuhara T, Taslima F, Abdullah M, et al. Probiotic Bifidobacterium breve prevents memory impairment through the reduction of both amyloid-β production and microglia activation in APP knock-in mouse. *J Alzheimers Dis.* 2022;85:1555–1571. doi: 10.3233/JAD-215025
24. Ben Haim L, Carrillo-de Sauvage M, Ceyzériat K, Escartin C. (2015). Elusive roles for reactive astrocytes in neurodegenerative diseases. *Front Cell Neurosci.* 9:278. doi: 10.3389/fncel.2015.00278
25. Trippier PC, Jansen Labby K, Hawker DD, Mataka JJ, Silverman RB. Target- and mechanism-based therapeutics for neurodegenerative diseases: strength in numbers. *J Med Chem.* 2013 Apr 25;56(8):3121–3147. doi: 10.1021/jm3015926. Epub 2013 Mar 27. PMID: 23458846; PMCID: PMC3637880.
26. Jewett BE, Thapa B. Physiology, NMDA Receptor. [Updated 2022 Dec 11]. In: *StatPearls* [Internet]. Treasure Island (FL): StatPearls Publishing; 2025 Jan-. Available from: https://www.ncbi.nlm.nih.gov/books/NBK519495/
27. Calabrese G, Molzahn C, Mayor T. Protein interaction networks in neurodegenerative diseases: from physiological function to aggregation. *J Biol Chem.* 2022 Jul;298(7):102062. doi: 10.1016/j.jbc.2022.102062. Epub 2022 May 25. PMID: 35623389; PMCID: PMC9234719.
28. Mohan Kumar D, Talwar P. Amyloid-β, tau, and α-synuclein protein interactomes as therapeutic targets in neurodegenerative diseases. *Cell Mol Neurobiol.* 2025 Oct 6;45(1):84. doi: 10.1007/s10571-025-01604-7. PMID: 41051722; PMCID: PMC12500485.
29. Karbalaei R, Allahyari M, Rezaei-Tavirani M, Asadzadeh-Aghdaei H, Zali MR. Protein-protein interaction analysis of Alzheimer's disease and NAFLD based on systems biology methods unhide common ancestor pathways. *Gastroenterol Hepatol Bed Bench.* 2018 Winter;11(1):27–33. PMID: 29564062; PMCID: PMC5849115.
30. Firdaus Z, Li X. Unraveling the genetic landscape of neurological disorders: insights into pathogenesis, techniques for variant identification, and therapeutic approaches. *Int J Mol Sci.* 2024 Feb 15;25(4):2320. doi: 10.3390/ijms25042320. PMID: 38396996; PMCID: PMC10889342.

31. Calderón-Garcidueñas L. Common fatal neurodegenerative diseases revisited: beyond age, comorbidities, and devastating terminal neuropathology there is hope with prevention. *Front Neurol.* 2022;13:901447. doi: 10.3389/fneur.2022.901447
32. Wang YA, Kammenga JE, Harvey SC. Genetic variation in neurodegenerative diseases and its accessibility in the model organism Caenorhabditis elegans. *Hum Genom.* 2017;11:12. doi: 10.1186/s40246-017-0108-4
33. Bocchetta M, Mega A, Bernardi L, Di Maria E, Benussi L, Binetti G, et al. Genetic counseling and testing for alzheimer's disease and frontotemporal lobar degeneration: an Italian consensus protocol. *J Alzheimer's Dis.* 2016;51:277–291. doi: 10.3233/JAD-150849
34. Vicidomini C, Fontanella F, D'Alessandro T, Roviello GN. A survey on computational methods in drug discovery for neurodegenerative diseases. *Biomolecules.* 2024 Oct 19;14(10):1330. doi: 10.3390/biom14101330. PMID: 39456263; PMCID: PMC11506269.
35. Marasco D, Vicidomini C, Krupa P, Cioffi F, Huy PDQ, Li MS, et al. Plant isoquinoline alkaloids as potential neurodrugs: a comparative study of the effects of benzo[c]phenanthridine and berberine-based compounds on β-amyloid aggregation. *Chem Biol Interact.* 2021;334:109300. doi: 10.1016/j.cbi.2020.109300
36. Caberlotto L, Nguyen TP, Lauria M, Priami C, Rimondini R, Maioli S, et al. Cross-disease analysis of Alzheimer's disease and type-2 diabetes highlights the role of autophagy in the pathophysiology of two highly comorbid diseases. *Sci Rep.* 2019;9:3965. doi: 10.1038/s41598-019-39828-5
37. Hong J, Luo Y, Zhang Y, Ying J, Xue W, Xie T, et al. Protein functional annotation of simultaneously improved stability, accuracy and false discovery rate achieved by a sequence-based deep learning. *Brief Bioinform.* 2020 Jul 15;21(4):1437–1447. doi: 10.1093/bib/bbz081. PMID: 31504150; PMCID: PMC7412958.
38. Jain A, Kihara D. Phylo-PFP: improved automated protein function prediction using phylogenetic distance of distantly related sequences. *Bioinformatics.* 2019;35:753–759.
39. Mateos A, Dopazo J, Jansen R, Tu Y, Gerstein M, Stolovitzky G. Systematic learning of gene functional classes from DNA array expression data by using multilayer perceptions. *Genome Res.* 2002;12:1703–1715.
40. Nariai N, Kolaczyk ED, Kasif S. Probabilistic protein function prediction from heterogeneous genome-wide data. *PLoS One.* 2007;2:e337.
41. Pearson WR. Protein function prediction: problems and pitfalls. *Curr Protoc Bioinformatics.* 2015;51:4.12.1–8.
42. Parikshak NN, Gandal MJ, Geschwind DH. Systems biology and gene networks in neurodevelopmental and neurodegenerative disorders. *Nat Rev Genet.* 2015 Aug;16(8):441–458. doi: 10.1038/nrg3934. Epub 2015 Jul 7. PMID: 26149713; PMCID: PMC4699316.
43. Casey BJ, Craddock N, Cuthbert BN, Hyman SE, Lee FS, Ressler KJ. DSM-5 and RDoC: progress in psychiatry research? *Nat Rev Neurosci.* 2013;14:810–814. doi: 10.1038/nrn3621

44. Winden KD, Oldham MC, Mirnics K, Ebert PJ, Swan CH, Levitt P, et al. The organization of the transcriptional network in specific neuronal classes. *Mol Syst Biol.* 2009;5:291. doi: 10.1038/msb.2009.46
45. Choi JK, Yu U, Yoo OJ, Kim S. Differential coexpression analysis using microarray data and its application to human cancer. *Bioinformatics.* 2005;21:4348–4355. doi: 10.1093/bioinformatics/bti722
46. Ballouz S, Verleyen W, Gillis J. Guidance for RNA-seq co-expression network construction and analysis: safety in numbers. *Bioinformatics.* 2015. doi: 10.1093/bioinformatics/btv118
47. Zhang B, Horvath S. A general framework for weighted gene co-expression network analysis. *Stat Appl Genet Mol Biol.* 2005;4:17. doi: 10.2202/1544-6115.1128
48. Zoppoli P, Morganella S, Ceccarelli M. TimeDelay-ARACNE: reverse engineering of gene networks from time-course data by an information theoretic approach. *BMC Bioinformatics.* 2010;11:154. doi: 10.1186/1471-2105-11-154
49. Lubs HA, Stevenson RE, Schwartz CE. Fragile X and X-linked intellectual disability: four decades of discovery. *Am J Hum Genet.* 2012;90:579–590. doi: 10.1016/j.ajhg.2012.02.018
50. Sliwoski G, Kothiwale S, Meiler J, Lowe EW Jr. Computational methods in drug discovery. *Pharmacol Rev.* 2013 Dec 31;66(1):334–395. doi: 10.1124/pr.112.007336. PMID: 24381236; PMCID: PMC3880464.
51. Ahlström MM, Ridderström M, Zamora I, Luthman K. CYP2C9 structure-metabolism relationships: optimizing the metabolic stability of COX-2 inhibitors. *J Med Chem.* 2007;50:4444–4452.
52. Amaro RE, Schnaufer A, Interthal H, Hol W, Stuart KD, McCammon JA. Discovery of drug-like inhibitors of an essential RNA-editing ligase in Trypanosoma brucei. *Proc Natl Acad Sci USA.* 2008;105:17278–17283.
53. Armour D, de Groot MJ, Edwards M, Perros M, Price DA, Stammen BL, et al. The discovery of CCR5 receptor antagonists for the treatment of HIV infection: hit-to-lead studies. *ChemMedChem.* 2006;1:706–709.
54. Baurin N, Baker R, Richardson C, Chen I, Foloppe N, Potter A, et al. Drug-like annotation and duplicate analysis of a 23-supplier chemical database totalling 2.7 million compounds. *J Chem Inf Comput Sci.* 2004;44:643–651.
55. Blum LC, Reymond JL. 970 million druglike small molecules for virtual screening in the chemical universe database GDB-13. *J Am Chem Soc.* 2009;131:8732–8733.
56. Euer N, Schwirzke M, Evtimova V, Burtscher H, Jarsch M, Tarin D, et al. Identification of genes associated with metastasis of mammary carcinoma in metastatic versus non-metastatic cell lines. *Anticancer Res.* 2002;22 (Suppl. 2A):733–740.
57. Huang N, Shoichet BK, Irwin JJ. Benchmarking sets for molecular docking. *J Med Chem.* 2006;49:6789–6801.
58. Hollingsworth SA, Dror RO. Molecular dynamics simulation for all. *Neuron.* 2018 Sep 19;99(6):1129–1143. doi: 10.1016/j.neuron.2018.08.011. PMID: 30236283; PMCID: PMC6209097.
59. Arkin IT, Xu H, Jensen MØ, Arbely E, Bennett ER, Bowers KJ, et al. Mechanism of Na^{+}/H^{+} antiporting. *Science.* 2007;317:799–803.

60. Borhani DW, Shaw DE. The future of molecular dynamics simulations in drug discovery. *J Comput Aided Mol Des.* 2012;26, 15–26.
61. Acharya C, Coop A, Polli JE, Mackerell AD Jr. Recent advances in ligand-based drug design: relevance and utility of the conformationally sampled pharmacophore approach. *Curr Comput Aided Drug Des.* 2011;7:10–22.
62. Alguacil LF, Pérez-García C. Histamine H3 receptor: a potential drug target for the treatment of central nervous system disorders. *Curr Drug Targets CNS Neurol Disord.* 2003;2:303–313.
63. Schneider G. Modeling Structure-Activity Relationships. In: *Madame Curie Bioscience Database* [Internet]. Austin (TX): Landes Bioscience; 2000–2013. Available from: https://www.ncbi.nlm.nih.gov/books/NBK6353
64. Shoichet BK. Virtual screening of chemical libraries. *Nature.* 2004 Dec 16;432(7019):862–865. doi: 10.1038/nature03197. PMID: 15602552; PMCID: PMC1360234.
65. Rizzo R, Wang D, Tirado-Rives J, Jorgensen W. Validation of a model for the complex of HIV-1 reverse transcriptase with sustiva through computation of resistance profiles. *J Am Chem Soc.* 2000;122:12898–12900.
66. Shoichet BK, McGovern SL, Wei BI, Irwin JJ. Lead discovery using molecular docking. *Curr Opin Chem Biol.* 2002;6:439–446. doi: 10.1016/s1367–5931(02)00339-3.
67. Paiva AM, Vanderwall DE, Blanchard JS, Kozarich JW, Williamson JM, Kelly TM. Inhibitors of dihydrodipicolinate reductase, a key enzyme of the diaminopimelate pathway of Mycobacterium tuberculosis. *Biochim Biophys Acta.* 2001;1545:67–77. doi: 10.1016/s0167–4838(00)00262-4.
68. Wei BQ, Baase WA, Weaver LH, Matthews BW, Shoichet BK. A model binding site for testing scoring functions in molecular docking. *J Mol Biol.* 2002;322:339–355. doi: 10.1016/s0022–2836(02)00777-5
69. Paul D, Sanap G, Shenoy S, Kalyane D, Kalia K, Tekade RK. Artificial intelligence in drug discovery and development. *Drug Discov Today.* 2021 Jan;26(1):80–93. doi: 10.1016/j.drudis.2020.10.010. Epub 2020 Oct 21. PMID: 33099022; PMCID: PMC7577280.
70. Lamberti MJ. A study on the application and use of artificial intelligence to support drug development. *Clin Ther.* 2019;41:1414–1426. doi: 10.1016/j.clinthera.2019.05.018
71. Kalyane D. Artificial Intelligence in the Pharmaceutical Sector: Current Scene and Future Prospect. In: Tekade Rakesh K, editor. *The Future of Pharmaceutical Product Development and Research.* Elsevier; 2020. pp. 73–107.
72. Baronzio G. Overview of methods for overcoming hindrance to drug delivery to tumors, with special attention to tumor interstitial fluid. *Front Oncol.* 2015;5:165. doi: 10.3389/fonc.2015.00165
73. Taye MM. Understanding of machine learning with deep learning: architectures, workflow, applications and future directions. *Computers.* 2023; 12(5):91. doi: 10.3390/computers12050091
74. Karhunen J, Raiko T, Cho K. Unsupervised Deep Learning: A Short Review. In: Bingham E, Kaski S, Laaksonen J, Lampinen J, Oja E, editors. *Advances in Independent Component Analysis and Learning Machines.* Academic Press: Cambridge, MA; 2015. pp. 125–142.

75. Deep Learning Techniques: An Overview. SpringerLink. Available online: https://link.springer.com/chapter/10.1007/978-981-15-3383-9_54 (accessed on 11 March 2023).
76. Keramida P, Syrigos NK, Kouvela M, Poulakou G, Charpidou A, Fiste O. AI-driven drug repurposing: applications and challenges. *Medicines (Basel).* 2025 Nov 13;12(4):28. doi: 10.3390/medicines12040028. PMID: 41283338; PMCID: PMC12641654.
77. Singh N, Halliday AC, Thomas JM, Kuznetsova OV, Baldwin R, Woon EC, et al. A safe lithium mimetic for bipolar disorder. *Nat Commun.* 2013;4:1332. doi: 10.1038/ncomms2320
78. Ghandikota SK, Jegga AG. Application of artificial intelligence and machine learning in drug repurposing. *Prog Mol Biol Transl Sci.* 2024;205:171–211. doi: 10.1016/bs.pmbts.2024.03.030
79. Trøseid M, Arribas JR, Assoumou L, Holten AR, Poissy J, Terzić V, et al. Efficacy and safety of baricitinib in hospitalized adults with severe or critical COVID-19 (Bari-SolidAct): a randomised, double-blind, placebo-controlled phase 3 trial. *Crit Care.* 2023;27:9. doi: 10.1186/s13054-022-04205-8
80. Meli R, Morris GM, Biggin PC. Scoring functions for protein-ligand binding affinity prediction using structure-based deep learning: a review. *Front Bioinform.* 2022 Jun 17;2:885983. doi: 10.3389/fbinf.2022.885983. PMID: 36187180; PMCID: PMC7613667.
81. Ain QU, Aleksandrova A, Roessler FD, Ballester PJ. Machine-learning scoring functions to improve structure-based binding affinity prediction and virtual screening. *Wiley Interdiscip Rev Comput Mol Sci.* 2015;5:405–424. doi: 10.1002/wcms.1225
82. Aldeghi M, Bluck JP, Biggin PC. Absolute Alchemical Free Energy Calculations for Ligand Binding: A Beginner's Guide. In: *Methods in Molecular Biology.* Springer: New York; 2018. pp. 199–232. doi: 10.1007/978-1-4939-7756-7_11
83. Bartók AP, De S, Poelking C, Bernstein N, Kermode JR, Csányi G, et al. Machine learning unifies the modeling of materials and molecules. *Sci Adv.* 2017;3:e1701816. doi: 10.1126/sciadv.1701816
84. Clark JJ, Benson ML, Smith RD, Carlson HA. Inherent versus induced protein flexibility: comparisons within and between apo and holo structures. *PLOS Comput Biol.* 2019;15:e1006705. doi: 10.1371/journal.pcbi.1006705
85. de Magalhães CS, Almeida DM, Barbosa HJC, Dardenne LE. A dynamic niching genetic algorithm strategy for docking highly flexible ligands. *Inf Sci.* 2014;289:206–224. doi: 10.1016/j.ins.2014.08.002
86. Gal Y, Ghahramani Z. Dropout as a Bayesian approximation: representing model uncertainty in deep learning. *PMLR.* 2016;48:1050–1059.
87. Gaulton A, Bellis LJ, Bento AP, Chambers J, Davies M, Hersey A, et al. ChEMBL: a large-scale bioactivity database for drug discovery. *Nucleic Acids Res.* 2011;40:D1100–D1107. doi: 10.1093/nar/gkr777
88. Huang DZ, Baber JC, Bahmanyar SS. The challenges of generalizability in artificial intelligence for ADME/Tox endpoint and activity prediction. *Expert Opin Drug Discov.* 2021;16:1045–1056. doi: 10.1080/17460441.2021.1901685

89. Imrie F, Bradley AR, van der Schaar M, Deane CM. Protein family-specific models using deep neural networks and transfer learning improve virtual screening and highlight the need for more data. *J Chem Inf Model.* 2018;58:2319–2330. doi: 10.1021/acs.jcim.8b00350
90. Jiang H, Fan M, Wang J, Sarma A, Mohanty S, Dokholyan NV, et al. Guiding conventional protein-ligand docking software with convolutional neural networks. *J Chem Inf Model.* 2020;60:4594–4602. doi: 10.1021/acs.jcim.0c00542
91. Koes DR, Baumgartner MP, Camacho CJ. Lessons learned in empirical scoring with smina from the CSAR 2011 benchmarking exercise. *J Chem Inf Model.* 2013;53:1893–1904. doi: 10.1021/ci300604z
92. Li Y, Liu Z, Li J, Han L, Liu J, Zhao Z, et al. Comparative assessment of scoring functions on an updated benchmark: 1. Compilation of the test set. *J Chem Inf Model.* 2014;54:1700–1716. doi: 10.1021/ci500080q
93. McNutt AT, Francoeur P, Aggarwal R, Masuda T, Meli R, Ragoza M, et al. GNINA 1.0: molecular docking with deep learning. *J Cheminform.* 2021;13:1–20. doi: 10.1186/s13321-021-00522-2
94. Liu J, Gui Y, Rao J, Sun J, Wang G, Ren Q, et al. *In silico* off-target profiling for enhanced drug safety assessment. *Acta Pharm Sin B.* 2024 Jul;14(7):2927–2941. doi: 10.1016/j.apsb.2024.03.002. Epub 2024 Mar 6. PMID: 39027254; PMCID: PMC11252485.
95. Aronson JK, Ferner RE. Joining the DoTS: new approach to classifying adverse drug reactions. *BMJ.* 2003;327:1222. doi: 10.1136/bmj.327.7425.1222
96. Lynch JJ 3rd, Van Vleet TR, Mittelstadt SW, Blomme EAG. Potential functional and pathological side effects related to off-target pharmacological activity. *J Pharmacol Toxicol Methods.* 2017;87:108–126. doi: 10.1016/j.vascn.2017.02.020
97. Mayr A, Klambauer G, Unterthiner T, Steijaert M, Wegner JK, Ceulemans H, et al. Large-scale comparison of machine learning methods for drug target prediction on ChEMBL. *Chem Sci.* 2018;9:5441–5451. doi: 10.1039/c8sc00148k
98. Perez-Nueno VI, Souchet M, Karaboga AS, Ritchie DW. GESSE: predicting drug side effects from drug–target relationships. *J Chem Inf Model.* 2015;55:1804–1823. doi: 10.1021/acs.jcim.5b00120
99. Zhang W, Liu F, Luo L, Zhang J. Predicting drug side effects by multi-label learning and ensemble learning. *BMC Bioinf.* 2015;16:365. doi: 10.1186/s12859-015-0774-y
100. Allen TEH, Goodman JM, Gutsell S, Russell PJ. Using 2D structural alerts to define chemical categories for molecular initiating events. *Toxicol Sci.* 2018;165:213–223. doi: 10.1093/toxsci/kfy144
101. Nguyen DA, Nguyen CH, Mamitsuka H. A survey on adverse drug reaction studies: data, tasks and machine learning methods. *Briefings Bioinf.* 2021;22:164–177. doi: 10.1093/bib/bbz140
102. Oprea TI, Bologa CG, Brunak S, Campbell A, Gan GN, Gaulton A, et al. Unexplored therapeutic opportunities in the human genome. *Nat Rev Drug Discov.* 2018;17:317–332. doi: 10.1038/nrd.2018.14

103. Chen D, Gao K, Nguyen DD, Chen X, Jiang Y, Wei GW, et al. Algebraic graph-assisted bidirectional transformers for molecular property prediction. *Nat Commun.* 2021;12:3521. doi: 10.1038/s41467-021-23720-w
104. Xiong Z, Wang D, Liu X, Zhong F, Wan X, Li X, et al. Pushing the boundaries of molecular representation for drug discovery with the graph attention mechanism. *J Med Chem.* 2020;63:8749–8760. doi: 10.1021/acs.jmedchem.9b00959
105. Hauser AS, Attwood MM, Rask-Andersen M, Schioth HB, Gloriam DE. Trends in GPCR drug discovery: new agents, targets and indications. *Nat Rev Drug Discov.* 2017;16:829–842. doi: 10.1038/nrd.2017.178
106. Wu Z, Ramsundar B, Feinberg EN, Gomes J, Geniesse C, Pappu AS, et al. MoleculeNet: a benchmark for molecular machine learning. *Chem Sci.* 2018;9:513–530. doi: 10.1039/c7sc02664a
107. Lake CR, Gallant S, Masson E, Miller P. Adverse drug effects attributed to phenylpropanolamine: a review of 142 case reports. *Am J Med.* 1990;89:195–208. doi: 10.1016/0002–9343(90)90299-s
108. Hutcheson JD, Setola V, Roth BL, Merryman WD. Serotonin receptors and heart valve disease: it was meant 2B. *Pharmacol Ther.* 2011;132:146–157. doi: 10.1016/j.pharmthera.2011.03.008
109. Ravna AW, Jaronczyk M, Sylte I. A homology model of SERT based on the LeuT(Aa) template. *Bioorg Med Chem Lett.* 2006;16:5594–5597. doi: 10.1016/j.bmcl.2006.08.028
110. Theodoris CV, Xiao L, Chopra A, Chaffin MD, Al Sayed ZR, Hill MC, et al. Transfer learning enables predictions in network biology. *Nature.* 2023;618:616–624. doi: 10.1038/s41586-023-06139-9
111. Selkoe DJ, Lansbury PJ Jr. Biochemistry of Alzheimer's and Prion Diseases. In: Siegel GJ, Agranoff BW, Albers RW, Fisher SK, Uhler MD, editors. *Basic Neurochemistry: Molecular, Cellular and Medical Aspects.* 6th ed. Philadelphia: Lippincott-Raven; 1999. pp. 949–968.
112. Kisby B, Jarrell JT, Agar ME, Cohen DS, Rosin ER, Cahill CM, et al. Alzheimer's disease and its potential alternative therapeutics. *J Alzheimers Dis Parkinsonism.* 2019;9(5):477. doi: 10.4172/2161–0460.1000477. Epub 2019 Sep 13. PMID: 31588368; PMCID: PMC6777730.
113. Matthews FE, Brayne C, Lowe J, McKeith I, Wharton SB, Ince P. Epidemiological pathology of dementia: attributable-risks at death in the Medical Research Council Cognitive Function and Ageing Study. *PLoS Med.* 2009;6: e1000180.
114. Heikkinen R, Malm T, Heikkila J, Muona A, Tanila H, Koistinaho M, et al. Susceptibility to focal and global brain ischemia of Alzheimer mice displaying abeta deposits: effect of immunoglobulin. *Aging Dis.* 2004;5:76–87.
115. Ohta S, Ohsawa I. Dysfunction of mitochondria and oxidative stress in the pathogenesis of Alzheimer's disease: on defects in the cytochrome c oxidase complex and aldehyde detoxification. *J Alzheimers Dis.* 2006;9:155–166.
116. Ali T, Yoon GH, Shah SA, Lee HY, Kim MO. Osmotin attenuates amyloid beta-induced memory impairment, tau phosphorylation and neurodegeneration in the mouse hippocampus. *Sci Rep.* 2015;5:11708.

117. Saburova EA, Vasiliev AN, Kravtsova VV, Ryabova EV, Zefirov AL, Bolshakova OI, et al. Human APP gene expression alters active zone distribution and spontaneous neurotransmitter release at the drosophila larval neuromuscular junction. *Neural Plast.* 2017;2017:9202584.
118. Rogers JT, Randall JD, Cahill CM, Eder PS, Huang X, Gunshin H, et al. An iron-responsive element type II in the 5′-untranslated region of the Alzheimer's amyloid precursor protein transcript. *J Biol Chem.* 2002;277:45518–45528.
119. Tamaoka A, Odaka A, Ishibashi Y, Usami M, Sahara N, Suzuki N, et al. APP717 missense mutation affects the ratio of amyloid beta protein species (A beta 1–42/43 and a beta 1–40) in familial Alzheimer's disease brain. *J Biol Chem.* 1994;269:32721–32724.
120. Cheng F, Sha Z, Zhou Y, Hou Y, Zhang P, Pieper AA, et al. The role for artificial intelligence in identifying combination therapies for Alzheimer's disease. *J Prev Alzheimers Dis.* 2025 Dec;12(10):100366. doi: 10.1016/j.tjpad.2025.100366. Epub 2025 Oct 27. PMID: 41145341; PMCID: PMC12627895.
121. Dantoine T, Auriacombe S, Sarazin M, Becker H, Pere J-J, Bourdeix I. Rivastigmine monotherapy and combination therapy with memantine in patients with moderately severe Alzheimer's disease who failed to benefit from previous cholinesterase inhibitor treatment. *Int J Clin Pr.* 2006;60(1):110–118. doi: 10.1111/j.1368–5031.2005.00769.x
122. Fang J, Zhang P, Zhou Y, Chiang CW, Tan J, Hou Y, et al. Endophenotype-based in silico network medicine discovery combined with insurance record data mining identifies sildenafil as a candidate drug for Alzheimer's disease. *Nat Aging.* 2021;1(12):1175–1188. doi: 10.1038/s43587-021-00138-z
123. Hassan YM, Wanas A, Ali AA, El-Sayed WM. Integrating artificial intelligence with nanodiagnostics for early detection and precision management of neurodegenerative diseases. *J Nanobiotechnol.* 2025 Oct 13;23(1):668. doi: 10.1186/s12951-025-03719-x. PMID: 41084041; PMCID: PMC12519733.
124. Palaniyappan R, Arthanari S, Subramanian M, Periyasamy P, Pethappachetty P, Ganesan R, et al. Nanotechnology-driven approaches in overcoming drug delivery challenges for neurodegenerative diseases. *Indian J Pharm Educ Res.* 2024;58(4s):s1185–s1200.
125. He S, Abarrategi JS, Bediaga H, Arrasate S, González-Díaz H. On the additive artificial intelligence-based discovery of nanoparticle neurodegenerative disease drug delivery systems. *Beilstein J Nanotechnol.* 2024;15(1):535–555.
126. Parra-Torres V, Melgar-Rodríguez S, Muñoz-Manríquez C, Sanhueza B, Cafferata EA, Paula-Lima AC, et al. Periodontal bacteria in the brain—implication for Alzheimer's disease: a systematic review. *Oral Dis.* 2023;29(1):21–28.
127. Vázquez-Mena O, Flores-Vázquez V, Gutiérrez-Sánchez E, Flores-Vázquez S, Gómez-Díaz S. A review of commercial and non-commercial wearables devices for monitoring motor impairments caused by neurodegenerative diseases. *Biosensors.* 2023;13(1):72.
128. Cho K, Wang X, Nie S, Chen ZG, Shin DM. Responsive theranostic systems: integration of diagnostic imaging agents and responsive controlled release drug delivery carriers. *Expert Opin Drug Deliv.* 2011;8(2):247–262.

129. Makhouri FR, Ghasemi JB. In silico studies in drug research against neurodegenerative diseases. *Curr Neuropharmacol.* 2018;16(6):664–725. doi: 10.2174/1570159X15666170823095628. PMID: 28831921; PMCID: PMC6080098.
130. Koutsoukas A, Simms B, Kirchmair J, Bond PJ, Whitmore AV, Zimmer S, et al. From in silico target prediction to multi-target drug design: current databases, methods and applications. *J Proteomics.* 2011;74:2554–2574.

Index